MW00571223

THE SOCIETY *of*
AMERICAN ARCHIVISTS

# aca

**Association of Canadian Archivists**

# Diplomatics

## *New Uses for an Old Science*

Luciana Duranti

Society of American Archivists
and
Association of Canadian Archivists

*in association with*

The Scarecrow Press, Inc.
Lanham, Maryland, and London
1998

Chapters 1 through 6 in this book were previously published as a series in *Archivaria,* appearing in the following issues: Chapter 1, Summer 1989; Chapter 2, Winter 1989-90; Chapter 3, Summer 1990; Chapter 4, Winter 1990-91; Chapter 5, Summer 1991; and Chapter 6, Winter 1991-92.

## SCARECROW PRESS, INC.

Published in the United States of America
by Scarecrow Press, Inc.
4501 Forbes Boulevard, Suite 200
Lanham, Maryland 20706

4 Pleydell Gardens, Folkestone
Kent CT20 2DN, England

British Library Cataloguing in Publication Information Available

**Library of Congress Cataloging-in-Publication Data**

Duranti, Luciana.
    Diplomatics : new uses for an old science / Luciana Duranti.
        p.    cm.
    Includes bibliographical references and idex.
    ISBN 0-8108-3528-2 (alk. paper)
    1. Diplomatics.    I. Title.
CD46.D87        1998
001.4'32—dc21                                    98-43004
                                                 CIP

♾™ The paper used in this publication meets the minimum requirements of American National Standard for Information Sciences—Permanence of Paper for Printed Library Materials, ANSI Z39.48–1984.
Manufactured in the United States of America.

# Contents

# Foreword

This book comprises a series of six articles first published consecutively in numbers 28 to 33 of *Archivaria*, the journal of the Association of Canadian Archivists, under the title "Diplomatics: New Uses for an Old Science." The author has added a lengthy introduction explaining her intellectual journey in writing the original articles. They are brought together here to provide archivists interested in diplomatics and its application to analysis of records of all kinds in any era with a convenient text on a subject little known outside the European tradition.

In the English speaking world, the literature on diplomatics is exclusively about special diplomatics, that is, on the application of diplomatics to the analysis of particular archival fonds, classes of archival documents in some period, or individual archival documents, often with the aim of historical understanding of them. This book is on the subject of general diplomatics, that is, on the concepts of diplomatic analysis in general terms. No text on general diplomatics has been available in English before its publication first as the series of articles and now in this single volume.

I had a small part in the genesis of this book, for I encouraged Professor Duranti to teach diplomatics to students in the Master of Archival Studies Program at the School of Library, Archival and Information Studies at the University of British Columbia in the hope that it would strengthen their intellectual armour. However, I had no idea how much it would contribute to the development of archival studies. In her introduction, the author explains how her students took to the subject with considerable enthusiasm and inventiveness, and how it provided them with the conceptual framework for several theses. Eventually learning general diplomatics found its way into the fundamental course all students take on the nature of archival materials. It now seems to me that no archivist can claim to understand archival documents without appreciating those concepts and being able to apply them in all kinds of situations.

Michel Duchein once wrote that "the archival document takes place at the heart of a functional process." If you want to understand how archival documents come into being, the form they take, their functional nature, and much more, I recommend you read this book.

Terry Eastwood
March 20, 1997

# Acknowledgments

This article is the last of the series on diplomatics, which began in *Archivaria* 28 (Summer 1989). Over the past three years many colleagues have generously offered their support, advice, ideas, and expertise to this illustration and exploration of Diplomatics for the modern archivist. It is with immense gratitude that I take the opportunity to acknowledge those contributions. My most heartfelt thanks go to Peter Robertson, previous General Editor of *Archivaria*, who believed in this project and made it possible; Terry Cook, who accompanied this effort with criticism and praise, challenge and encouragement, step by step, assiduously and enthusiastically all the way; Terry Eastwood, who first had the idea of introducing Diplomatics into the M.A.S. curriculum at U.B.C., encouraged me to write about it, and constantly helped me to gain a North American perspective on a discipline so deeply rooted in the European tradition; Ronald Hagler, who, with infinite patience, taught me the secrets of a clear English style and many "good English words"; Hugh Taylor, who continually offered support and suggestions; David Bearman, who contributed stimulating ideas and enthusiastic interest; Charles Dollar, who played devil's advocate all the time; John McDonald, who with his experience confirmed my intuitions; and Maynard Brichford, Helen Samuels, Margaret Hedstrom and all those readers who spontaneously offered opinions and suggestions. Finally, special thanks go to all the M.A.S. students at U.B.C. who took the course in Diplomatics, particularly those who chose to continue the study of diplomatic concepts and methods in their theses, for having provided constant inspiration and stimuli; and to all the archivists who participated in the workshops which I offered, for their act of faith, their intellectual curiosity, and their inspiring observations and questions. If Diplomatics as exposition of doctrine will disappear from *Archivaria* with the present article, I hope that Diplomatics as a body of concepts informing the thinking and writing of professional archivists will be more and more present in its pages, as the application of those concepts shows their validity and versatility.

# Introduction

I wrote this series of articles primarily for my Canadian students. As part of my archival education in Italy, I had accumulated four years of study of diplomatics, a subject embedded in the standard academic curriculum for archivists, both in the university and the state archives schools. I never knew why diplomatics was considered an essential component of archival education, as I was never told and never asked. I found out when I came to North America as an archival educator.

During my first year as a professor in the Master of Archival Studies (MAS) program at the University of British Columbia (UBC), I was responsible for three courses, a first year annual course illustrating all the archival concepts and principles affecting the traditional archival functions and the methods derived from them, a first year one term course on records management, and a second year one term course presenting the history of archives and archival science. In connection with the history of archival science, I found myself explaining to second year students the origin of the philological sciences and, in particular, the basic concepts, principles and methods of diplomatics.[1] This exercise was a journey of discovery for me as well as for the students.

---

1. The idea came from the Chairman of the MAS Program, Prof. Terry Eastwood. When he indicated that I should dedicate some time to the teaching of diplomatic concepts, I expressed surprise at the suggestion that Canadian students could have any use for a discipline concerned with medieval records. He insisted, adding that this would have also fulfilled the wishes of Roy Stokes, the director of the then School of Librarianship in the 1970s, who brought to fruition his own idea of offering a program of education specifically for archivists, which would include courses of diplomatics and paleography.

To explain the concepts of a science developed for the analysis and criticism of medieval documents to individuals who have never seen medieval documents and are not very familiar with medieval history may be regarded as a difficult task, but to me the real challenge was to make the students see the relevance of diplomatic study to their future professional work with records which could hardly be two centuries old, especially considering that I could not see it myself. I decided that I was not even going to try. Thus, I approached the subject academically rather than professionally. I told the students that they were going to acquire some diplomatic knowledge as part of their archival cultural baggage because it would enhance their understanding of the evolution of their discipline and of the ideas about records, and then I proceeded to expose diplomatic concepts and their application to documentary realities of the past. In other words, I treated the discipline as a "formative" subject.

I can still feel the sense of surprise and revelation that came over me when the students began to interrupt my illustrations in order to make connections with the contemporary reality, to attempt applications of concepts to modern records, to use the old methods to address present issues, to announce their sudden understanding of things learned earlier in their study of archival functions, to express their excitement at the discovery of a consistent and rigorous system of ideas to which they could refer when confused, and to ask for texts of diplomatics they could read.

At the same time, I was feeling rather frustrated in the teaching of both first year courses. I found it very difficult to explain the nature of an archival fonds and of the aggregations composing it without analysing its elementary parts, the records. I found it even more difficult to teach records creation, maintenance and use, and how to accomplish archival functions without using concepts rooted in diplomatics. Thus, as soon as I saw the reaction of the second year students to the teaching of diplomatics, I began to introduce some diplomatic ideas in the first year courses, at first, tentatively, and then, after observing the effects, with increasing confidence. The first year students could easily absorb the concepts and base their understanding of archival science on them. Moreover, they found diplomatic notions so much more rigorous than records management notions and therefore so much more reliable as points of reference in their journey towards the acquisition of knowledge about records, that they too began to ask for readings of diplomatics.

Readings? Readings in English? Diplomatics readings in English were written either by medieval historians, with a view to the analysis of the sources for historical interpretation, or by diplomatists/archivists, with a view to the identification of a specimen or to the illustration of a particularly significant group or type of records. English writers are not inclined to immerse themselves in diplomatic theory, or "general diplomatics," and prefer to apply accepted concepts and principles to concrete realities for mostly practical purposes. In contrast, French and German writers tend to articulate and explore diplomatic ideas and their implications. However, both ideas and implications are demonstrated by examples taken from French and German chanceries operating many centuries ago. Students who are not familiar with those chanceries cannot appreciate the significance of the examples. Italian writers of the last twenty years have attempted to extend the meaning of diplomatic concepts to embrace modern records, and to develop them to deal with modern bureaucracy. The content of their work could be accessible to Canadian students, but certainly not the language.

What could I possibly do about the problem of lack of texts in English capable of supporting my class instruction? I could write a book of diplomatics for modern archivists. However, such an endeavour would take too long to be useful to my first five or six classes. Moreover, this kind of effort requires continuous feedback, as it can proceed only through trial and error, and the writing of a book is a solitary exercise, which does not allow for constant testing of ideas. Thus, the best solution appeared to be that of writing a book in installments, and I made a proposal to the then editor of *Archivaria*, Peter Robertson, for a series of six articles that would cover diplomatic concepts, principles and methods from the point of view of the contemporary archivists.

To me, the idea had several advantages: I could have something for the students to read almost immediately; I could receive comments about each article before sending the following one to publication, and this would allow me to address them in the proper context, without having to carry out a theoretical discussion in letters to editors; I could present my ideas and the ideas of my critics in class discussions thereby receiving precious feedback and stimulating the students' minds; I could rethink my hypotheses and adjust them in the final article in light of what I had learned in the process of writing the previous ones; and, by the end of such process I would have in my hands a whole compendium of diplomatic knowledge (where the footnotes were meant to serve as a basic bibliography) to give the students to study for their exams and to reflect upon in the following years as professionals. An added benefit was the opportunity to introduce my North American colleagues to a discipline largely considered irrelevant by archivists responsible for modern records, and to engage them in a debate which could produce useful results, if only by reason of being a debate about the nature and characteristics of archival material.

The decision of dividing general diplomatics into five parts and using a sixth article for the purpose of reviewing the main ideas and suggesting their possible use in the modern records environment was instinctive, as it did not derive from a carefully planned outline of the content of the articles, but from the sense I developed in the course of my teaching of the key concepts that diplomatic knowledge could contribute to the archival understanding of contemporary records, those of authenticity and originality, fact and act, persons, procedure, and form.

The first article was a real challenge from the language point of view. This was my first major effort at writing in English, and I had to wrestle with issues of composition (e.g., I had never written a sentence shorter than five lines and a paragraph shorter than one page), style (e.g., Italians use italics for titles of articles and quotation marks for titles of journals), wording (e.g., some cognates have a different meaning in Italian and English;[2] and some words sounded to me better than others, but were not the most commonly used[3]), and terms (e.g., some diplomatic terms, such as authentic

---

2. For example, transaction in Italian is "transazione," but it means "compromise" in common language and is a specific type of financial act in technical language.
3. For example, in my draft I kept using the word "exigency," and one of my readers noted on the margin that "need" is a perfectly good English word for what I meant to say.

and original, had meanings conflicting with common and legal usage).

However, from a content point of view, I did not encounter many problems, for two main reasons. Firstly, the nature of the introductory article dictated its development: why the series, what it was about, origin and evolution of its subject matter, the substance of the matter, etc.. Secondly, I was not attempting to recruit followers, or simply to persuade the readers that diplomatics was worth learning. I was responding to an immediate practical need, to provide archival students with a text to study, and to a perceived intellectual need, to explain for the North American archivists what diplomatics consisted of, what its purposes were, and how they developed over time. I was neither providing my personal interpretation of diplomatic ideas (although, one inevitably does), nor attempting to approach those ideas differently, in a creative way, and develop them in new directions. I was telling the diplomatic story as I understood it, and as concisely as possible, given the limited available space. The title I came up with, "Diplomatics: New Uses for an Old Science," was conceived only for this first article, and was meant to convey the simple message that the old science could be used in new ways. However, in order to do so, its fundamental concepts and principles had to be understood and absorbed into archival knowledge. The succeeding articles were going to describe those concepts and principles one by one, and their titles would refer to their specific content.

The explanation of diplomatic ideas in the first article follows the common international understanding. A case in point is the definition of "written document"[4] The document is defined as "evidence," as traditional diplomatists do, because diplomatics, since its origin, was meant to look at documents retrospectively, as source of proof of facts that needed to be demonstrated. However, the same diplomatists, when defining documents in relation to their nature, as determined at the time of their creation, rather than for purposes of diplomatic criticism, call them "instruments" (*in-*

---

4. "diplomatics studies *the written document,* that is, evidence which is produced on a medium...by means of a writing instrument...or of an apparatus for fixing data, images and/or voices," p. 41.

*strumenta*, means for carrying out actions), not "evidence."[5] This is why, in the same article, I do not use the term evidence when defining "archival documents."[6] In fact, two things need to be pointed out. The first is that diplomatics attributes the capacity of being used as evidence to *all* written documents, that is to any sort of recorded information, not just to archival documents or records.[7] The second is that diplomatics gives the term "evidence" a very specific meaning. Heather Mac Neil explains:

> The rules of textual criticism enunciated in Mabillon's treatise...reflect the new conception of evidence as inference....Diplomatic analysis translated a document into a system of external signs or traces which pointed to a reality beyond themselves....Moving from the observation of perceptible matters of fact (the elements of the document itself) to assertions about imperceptible matters of fact (the past in which the document was created), diplomatic methodology transformed written facts into historical sources and nurtured the belief that knowledge about a past to which there was no direct access could, nevertheless, be attained by examining its documentary traces.[8]

By the time the first article was available in print, several things had happened. The MAS program had gone through a major revision of the curriculum. As a member of the ACA Education Committee, I had participated in the drafting of the new ACA

---

5. The term evidence in general refers to the use that one makes of something. This is because evidence is a relationship between a fact to be proven and the fact that proves it. Thus, one uses the latter to ascertain the former. In the absence of a fact to be proven, there is no fact that proves it, there is no evidence. This simply means that the concept of evidence is at the same time much broader and much more specific than that of archival document or record.

6. "the object of diplomatics is not any written document it studies, but only the *archival document*, that is a document created or received by a physical or juridical person in the course of a practical activity," p. 42.

7. Throughout this introduction, the terms "archival document" and "record" are used interchangeably.

8. Heather MacNeil, "Assessing the Trustworthiness of Documentary Evidence: From the *Corpus Iuris Civilis* to *De Re Diplomatica*," paper delivered at "Issues of Evidence," a symposium sponsored by the Medieval and Renaissance Studies Interdisciplinary Group of Green College, University of British Columbia, 16 November 1996.

guidelines for graduate archival education.[9] I had shared the text of the manuscript of this first article with some North American colleagues and obtained much feedback. I had offered diplomatics workshops at the Association of Canadian Archivists (ACA) and Society of American Archivists (SAA) meetings and received more feedback. Some of my students had become interested in pursuing the analysis of diplomatic concepts in work situations.

The revision of the MAS curriculum was a natural consequence of growth in the number of instructors and students, and of the experience of seven years of dedicated archival education, which clearly indicated the need for a more complex and articulated system.[10] Part of the revision was inspired by the perspective brought by diplomatics. A new course entirely dedicated to diplomatics was designed for the second year students, and a new course on legal concepts and issues was conceived to provide support to the teaching of diplomatic concepts. The tight relationship between diplomatic and legal concepts had constituted a problem for me since the beginning, mainly because I was educated in a civil law environment and found it difficult to identify corresponding terms for the same concepts in the common law system in which I was teaching. In the first years of my Canadian experience, I spent innumerable hours studying law dictionaries and manuals of administrative law. I arrived at the conclusion that, while it was possible and indeed necessary to learn general diplomatics without making reference to any specific legal system, it was impossible to absorb its concepts without a solid understanding of legal thinking.[11] Even more, it was impossible to use them in new ways with

---

9. Education Committee, Association of Canadian Archivists, "Guidelines for the Development of a Two-Year Curriculum for a Master of Archival Studies Programme (December 1988)," *Archivaria* 29 (Winter 1989-1990): 128-141.

10. For a discussion of the changes to the MAS curriculum and their rationale, see Terry Eastwood, "Nurturing Archival Education in the University," *The American Archivist* 51 (Summer 1988): 228-252.

11. Legal thinking in the Western world is entirely rooted in Roman law and Roman jurisprudence. The *ius commune* of Medieval Europe was the framework of Roman legal concepts that superimposed itself over the *iura propria*, the individual laws of discrete national groups. Modern English jurisprudence is increasingly acknowledging the direct link between Roman law and the common law of England. See for example: John F. Winkler, "Roman Law in Anglo-Saxon England," *Journal of Legal History* 13, no. 2 (1992): 101-127, and Charles Donahue, Jr., "Ius Commune, Canon Law, and Common Law in England," *Tulane Law Review* 66 (1992): 1745-1780.

contemporary records without a good general knowledge of Canadian law.[12]

The drafting of the guidelines for graduate archival education gave me the opportunity of examining in some depth, together with Canadian archivists working in the field (i.e., the other members of the ACA Education Committee), the reasons behind the inclusion of diplomatics in all curricula of autonomous graduate degrees in archival studies existing at the time in Europe, South America and Africa.[13] At the same time, my colleagues in the Education Committee, having had the opportunity of reading the manuscript of my first article, had formed their own sense of the need for diplomatic knowledge, which was quite different from that which my students had manifested.

My students had instinctively regarded diplomatics as both a theoretical framework which allowed them to understand archival ideas and principles, and a rigorous system of concepts and norms that disciplined their search for consistent methods of handling contemporary records. In contrast, my colleagues in the Education Committee believed that the most important contribution of diplomatics to archival knowledge was its methodology of analysis: thus, it was the procedure of diplomatic criticism and its approach to the examination of records that warranted the inclusion of

---

12. Terry Eastwood explained the reasons for new courses in diplomatics and law as follows: "the composition of documents, how they are put together, is surely a valid study of archivists. Form, which of course follows function, is a surer immediate means of identification than the function behind records, which is more difficult to construe....So study of diplomatics instrumentally assists the basic functions of identification, arrangement, description, and appraisal. Legal knowledge is an absolutely vital and almost completely neglected area of study by archivists in North America....As documents arising from actions or preparing for actions (in administrative or private life) which have or may have legal significance (either directly or contextually), archives are suffused with and by law....As public officials (which all archivists are),...archivists need to know the nature of the law and its influence on the documentation process." Eastwood, "Nurturing Archival Education in the University," p. 248.

13. At the time, there were not autonomous self-contained archival graduate degrees in North America, other then at the University of British Columbia, or in other parts of the world; and, to my knowledge, still today, there are no such degrees in North America or in Asia or Australia, the only exception being the UBC MAS programme.

diplomatics among the subjects recommended by the committee in its ideal curriculum for graduate archival education.[14]

As to myself, I had arrived at the conclusion that the reasons for which traditional archival programmes in other parts of the world included diplomatics as a required subject of study (i.e., because diplomatics is considered both a formative discipline and a necessary instrument for dealing with medieval and early modern records) had to be disregarded in our deliberations, as our primary responsibility was to design a curriculum viable in the Canadian context and capable of addressing present and foreseeable Canadian needs. Moreover, some dose of realism had to be brought into committee decisions. While my students' perception was more convincing to me than that of my colleagues, I had to admit that to require the inclusion of diplomatics among the core subjects of any curriculum of archival education was perhaps unrealistic. Even if all Canadian archivists who had never heard of diplomatics before were to trust our wisdom and call for archival programs with a significant element of study of diplomatics (something which I strongly doubted), where could Canadian universities find archival instructors with enough diplomatic knowledge to deliver a core course on the subject?

In the meanwhile, I had collected a significant amount of response from both Canadian and American colleagues who had been asked to comment upon the manuscript of the first article and who had taken part in my diplomatics workshops. Everyone expressed interest, some manifested excitement, many were rather doubtful of the existence of a practical use for diplomatic knowledge, but nonetheless most loved the intellectual challenge posed by the rigour imposed of diplomatic criticism, and believed that I

14. Diplomatics was listed in the guidelines among the "subjects for method courses," a category of subjects included in the archival curriculum for the purpose of providing "the archivist with a variety of methodology for the intellectual and physical control of archival material." This was the rationale: "Whereas archival science addresses collectivities of archival documents, diplomatic criticism focuses on analysis of the formation, forms, and effects of single archival units (e.g., documents, volumes, registers). Study of the genesis, inner constitution, and transmission of documents illuminates the relationship between their context, content, and form which is at the heart of archival work." Education Committee, Association of Canadian Archivists, "Guidelines for the Development of a Two-Year Curriculum for a Master of Archival Studies Programme (December 1988)," 135, 138.

should continue in my endeavour. Quite in contrast with the attitude shown by my colleagues, my students found the intellectual
challenge quite exhausting, but were ready to pay the price, being
deeply persuaded that what they were learning was going to
facilitate their future archival work, provide clear and consistent
responses to problems that archivists seemed unable to solve in
practice, and contribute to the development of the archival discipline. The skeptical enthusiasm (I cannot find any better way of
describing it than an oxymoron) of my colleagues and the tired,
fatalistic confidence of my students changed completely my approach to the task. When I wrote the second article, I was no longer
explaining diplomatics: I was proselytizing!

The writing of the second article was extremely hard, partly,
even if not only, because of the different mood with which I was
approaching it. In order to persuade my readers of the relevance of
diplomatics for both modern archival discourse and work and the
North American environment, and to promote its study, rather
than simply illustrate its content, I needed to use as authoritative
support well known contemporary literature expressing either
consistent and complementary ideas, or concerns that could be
directly addressed by diplomatic concepts and methods. While
Gerald Gall's explanation of the Canadian legal system was useful
to demonstrate the concept of juridical system,[15] my greatest allies
were a book of sociology and an international report. The discovery
of Stanley Raffel's *Matters of Facts*[16] was a turning point in my
approach to diplomatics. Raffel's perspective on the relationship
between records and events, bureaucracy and records, and records'
authors and bureaucracy made me see how diplomatics could be
used as a standard guiding records creation rather than just assessing it. At the same time, the publication of the United Nations

---

15. Gerald Gall, *The Canadian Legal System*, 2nd ed. (Toronto, Calgary, Vancouver:
Carswell Legal Publications, 1983). Although, the authority of this author failed to
persuade some of my American colleagues, one of whom bluntly stated that America
does not have anything comparable to a juridical system.

16. Stanley Raffel, *Matters of Fact* (London, Boston, and Henley: Routledge and
Kegan Paul, 1979).

guidelines on the management of electronic records gave me the
unique opportunity to demonstrate the need for such a standard.[17]

The definition of records provided by this report, which inextricably linked them to administrative action and procedure; its attempt to systematise the available knowledge of records
management and of the products of electronic technology on the
basis of some overriding concepts and principles; its search for
methods of controlling electronic records which would be consistent with those required for traditional records; and its realization
that the quest for a disciplined approach to the management of
electronic records was proceeding by trial and error, in a vacuum
of theoretical understanding, offered the best possible stage for the
presentation of diplomatics as both the intellectual framework—
tested throughout the centuries in many different juridical systems—within which the problems presented by contemporary
records had to be addressed, and the body of knowledge that,
properly developed, would solve all such problems. Diplomatics
could fulfill these two functions by providing an internally consistent self-referential system of ideas, norms and methods capable of
dealing with traditional, as well as electronic, material, and by
constituting itself in a complex, rigorous, all encompassing standard acceptable to everyone because of its decontextualised nature.

The writing of this second article was harder for another reason,
somewhat connected to my new attitude towards my articles. It
was clear to me at the outset that any attempt to introduce a
rigorous diplomatic terminology, which is the foundation of the
entire diplomatic discourse, both theoretical and methodological,
had to take into account the inconsistency of the current archival
terminology in English and its historical development. If I were still
in the explanatory mood, I would probably have limited my effort
to making connections between diplomatic terms and the archival
terms I had learned from Jenkinson and the other British writers
whom I had studied in school, without going to great lengths to use

17. United Nations Advisory Committee for the Co-ordination of Information
Systems (ACCIS), Technical Panel on Electronic Records Management (TP/REM),
*Electronic records guidelines: a manual for policy development and implementation* (Ge-
neva and New York: United Nations, 1989).

the terms my readers were most familiar with, even when I considered them to be in contrast with the most basic notions of archival
theory. But, as I was in a proselytizing mood, I made a special effort
to be understood. For example, the four functional categories of
records (i.e., dispositive, probative, supporting, and narrative) are
discussed in relation to the terms *records* and *manuscripts*.[18] This
distinction, based on the nature of the will creating the archival
document and on the effects that it is meant to produce, does not
appear anywhere else in the diplomatic series, and certainly today
I would not have used it, as it confused the readers more than it
helped them.

Another concession I made to incorrect legal terminology, primarily because I wanted the reader to focus on the issue at hand,
the nature of organizational records, rather than on a marginal one,
the nature of organizational actions. In discussing the statement of
the United Nations' report that records are "recorded transactions," I used the term *transaction* with the same meaning in which
the report used it (i.e., any form of communication.)[19] However, in
consideration of the fact that diplomatics attributes the term transaction the same meaning given it by the law, and that, because of
my misjudgment, my students had taken the habit of using the term
improperly (it was very interesting to see how students would
absorb much faster what I wrote, including the language I used,
than what I said in class), I returned on the subject in the sixth article
of the diplomatics series and illustrated the true meaning of the
term, as used by diplomatists and archivists.[20]

The most important thing about the second article is that, for the
first time, and almost unconsciously, I began developing diplomatic theory. Probably, what caused it was again my new attitude
towards the series: if I wanted to show the applicability of diplomatic ideas to contemporary situation, I had to develop them until
they became applicable, if necessary. This is quite apparent in the

18. Pp. 69-70.
19. Pp. 73-74.
20. Pp. 169-170.

course of the discussion of the functional categories of records. The two categories identified by classic diplomatics, *dispositive* records and *probative* records, were simply inadequate for dealing with all kinds of records generated by modern creators, particularly when one started to consider electronic databases. Thus, I conceived of two more categories, *supporting* records and *narrative* records, which would accommodate all records that did not have a legal nature.[21]

This decision turned out to be very useful for the future developments of diplomatic analysis, especially when, in the course of a research project undertaken four years later, I had to identify and define the characteristics of electronic systems and their nature. However, it also provided significant support to the research that the students were conducting. A good example is represented by the thesis of Heather Heywood, which aimed to define legal value and identify its components, to distinguish between actual and potential legal value, to determine the strength of potential legal value, and to propose methods for assessing the legal value of records. Her thesis relied on the functional categorization of records as developed in the second article, which proved to be rigorous enough to enable her to address consistently the issues involved in the attribution of legal value, without running into obstacles *de iure* or *de facto*.[22]

After the publication of the second article, a couple of events contributed to stir the direction of my subsequent articles. First, I was awarded by the Association of Canadian Archivists the 1990 W. Kaye Lamb Prize for the first article of diplomatics. The citation read: "Duranti's article on diplomatics, the first of a series, brings to North American audiences the essential elements of an ancient European discipline, enabling the archival profession to reclaim an important part of its intellectual heritage. Her insistence that archivists must carefully study the form and structure of the documents in their care supports the research-based approach to archival work

---

21. P. 69.
22. Heather M. Heywood, "Appraising Legal Value: Concepts and Issues" (Master of Archival Studies Thesis, University of British Columbia, 1990).

which *Archivaria* has endeavoured to foster.''[23] This citation, even more than the prize itself, represented for me both a sanction and a new charge. As I saw it, my Canadian colleagues were accepting the discipline of diplomatics as part of their intellectual heritage, and wanted to be able to claim it as their own. To me, this meant that I was well on my way to accomplishing my initial aim, making diplomatics accessible to Canadian archivists by explaining the discipline and its purposes, and that now I was entrusted with the distinct responsibility of adapting diplomatic concepts and methods to Canadian needs. I regarded this as an awesome task. Fortunately, I was not left alone to accomplish it.

The second event that occurred at this time and had a significant influence on the choices I made was the fact that my students began to proselytize too. A few of them began to choose thesis topics centered on diplomatics, while many more began to incorporate diplomatic concepts and methods within theses focused on other areas of investigation.[24] One wrote an article explaining what

---

23. The W. Kaye Lamb Prize, established in honour of a past Dominion Archivist and National Librarian, is given annually to the author of an article appearing in *Archivaria*, the journal of the Association of Canadian Archivists, which by the quality of its research, reflection and writing most advances archival thinking and scholarship in Canada. The quoted citation is written on the original certificate and signed by Peter Robertson, then General Editor of the journal. The afternoon I spent with the then director of the School of Library, Archival and Information Studies, Basil Stuart-Stubbs, at the apartment of W. Kaye Lamb to toast the prize and diplomatics still constitutes one of my most cherished memories.

24. To date, the theses which entirely focus on diplomatics are: Steven Davidson, "The Registration of a Deed of Land in Ontario: A Study in Special Diplomatics," (Master of Archival Studies thesis, University of British Columbia, 1994); Anthony Gregson, "Records Management Attributes in International Open Document Exchange Standards" (Master of Archival Studies thesis, The University of British Columbia, 1995); Joni Mitchell, "Civil Litigation, Probate and Bankruptcy Procedures: A Diplomatic Examination of British Columbia Supreme Court Records" (Master of Archival Studies thesis, The University of British Columbia, 1995), Janice Simpson, "Broadcast Archives: A Diplomatic Examination," (Master of Archival Studies thesis, The University of British Columbia, 1994); Janet Turner, "Special Diplomatics and the Study of Authority in the United Church of Canada" (Master of Archival Studies, The University of British Columbia, 1994). The dates of graduation do not reflect the actual period of research and work on the theses. Examples of other theses which grounded many of their findings on diplomatic concepts are the already cited work on legal value, and Erwin Wodarczak's "The Facts About Fax: Facsimile Transmission and Archives" (Master of Archival Studies, The Univer-

diplomatics represented to her.[25] The support and enthusiasm of my students did much to reinvigorate my promotional effort and my belief that the possibility of developing diplomatics to the point of transforming it in an instrument fully and consistently usable by contemporary archivists of all nationalities was not far-fetched. Moreover, North American colleagues began to send me reproductions of records preserved in their institutions to use in class and for my research,[26] and to comment on my articles in a constructive way, by suggesting solutions to problems of application of concepts to modern situations as well as pointing out inconsistencies or gray areas. These contributions begin to appear in the third article, where all documents analysed and discussed are taken from a Canadian context, and where a discussion is initiated about possible adaptation of classic diplomatic ideas to a North American context.[27] In particular, the proposal submitted by a colleague that the public or private nature of archival documents be determined by the public or private nature of the person establishing the procedure in which they participate is fully described, and its adoption is supported.[28]

The fourth article shows very clearly that I had finally found my voice.[29] I was still explaining classic diplomatics, but I was doing so only to provide the conceptual background to the new ideas that I was introducing, and to show how the latter were consistent with the former and directly building on them. This is particularly evident in the development of the discussion related to procedural phases and types.[30] The primary consequence of my new "crea-

ity of British Columbia, 1991).

25. Janet Turner, "Experimenting with New Tools: Special Diplomatics and the Study of Authority in the United Church of Canada," *Archivaria* 30 (Summer 1990): 91-103.

26. Among these colleagues, the most constant has been Patricia Kennedy, from the National Archives of Canada, whose sensible choices, accompanied by transcriptions and explanations, have been very precious for my work.

27. Pp. 81-106.
28. Pp. 105-106.
29. Pp. 107-131.
30. Pp. 109-114.

tive" mood was that the ideas I was proposing were not as well developed as the classic ones, and at times were only raw intuitions, some of which elicited no response, while others found a large resonance in the archival community, evolved into complex concepts, and became the intellectual foundation for systematic research and practical applications. An example of the latter intuitions is represented by the statements that "an understanding of procedures [in diplomatic terms] is the key to the understanding of [electronic] information systems," and that, with electronic records, we should use the same diplomatic approach that we use with traditional records.[31]

A secondary consequence of my creative mood was that my new ideas continued to develop after having been published in the series of articles. Of course, classic concepts had evolved too, but their adaptation and change had happened in the past: I could illustrate the meanings attributed to the concepts overtime, and suggest possible ways of applying them to modern reality. Now, on the contrary, I was adapting the concepts as I was going along, in some cases without bothering to describe their original use, but I did not have enough time to develop them completely. This happened mostly in the fifth article.[32]

I had the clear sense that the traditional identification of the elements of documentary form, and their division into extrinsic and intrinsic could not work with post-medieval records. I thought it possible to address the issue and solve it by developing some of the elements, discarding others, and moving a few from one category to another. For example, records annotations obviously represented the most vital and explicit link between form and procedure, and needed to be brought to the forefront as a key element, to be categorised, and to be analysed in some depth. I went as far as I could in meeting such a need, given the limitations imposed by printing deadlines,[33] but I realised a year later that I did not go far

---

31. Pp. 130-131.

32. Pp. 133-158.

33. I had committed myself to submit an article every six months, because the series had to include six *consecutive* installments occupying the lead-article space in as many issues of the journal. When I began to develop my own diplomatic ideas,

enough: in the article, annotations are still presented as extrinsic elements, and this misrepresents their function.[34] In fact, because they constitute a bridge between the record and the actions that put it into existence, handle it in the course of procedure, and maintain it through time as a fact in itself, annotations are neither extrinsic or intrinsic elements of form. I now recognise that, with modern records, we encounter three major changes in their formal elements: 1) annotations are part of the intellectual form of the record, as opposed to its physical form;[35] 2) the traditional correspondence between intellectual form and intrinsic elements and physical form and extrinsic elements cannot be considered valid;[36] and 3) physical form does not include the medium among its components.[37]

The sixth article was initially meant to serve as a conclusion,[38] but I also used it as an opportunity to approach in a creative mood concepts that I had presented when I was in the explanatory

---

this became a major problem, because I had no time to do any substantive research to validate or expand them. To me, this is now painfully visible in the fifth and sixth articles, as a consequence of three years of research conducted on some of the specific concepts treated in them. Readers might be interested in comparing these articles with a forthcoming one discussing the same concepts as they have been developed in the past three years: Luciana Duranti and Heather MacNeil, "The Preservation of the Integrity of Electronic Records: An Overview of the UBC-MAS Research Project," *Archivaria* 42 (Fall 1996).

34. P. 140.

35. The intellectual form is the whole of the formal attributes of the record that represent and communicate the elements of the action in which the record is involved and of its immediate context, both documentary and administrative. The physical form is the whole of the formal attributes of the record that determine its external make-up. The elements of both intellectual and physical form are meant to convey determined types of meaning.

36. For example, for the reasons argued in the text, intellectual form must today include the annotations, which classic diplomatics considers extrinsic elements, on the grounds that they are added after the making or receiving of the document. In another example, classic diplomatics classified the medium among the extrinsic elements of form, as in medieval times the type of medium, the way it was treated and cut, and the way in which it supported the writing were meaningful and meant to contribute to the message conveyed by the intrinsic elements of form. But, today, the medium cannot be considered a part of the physical form.

37. The medium is not part of the physical form because, in modern records, it is not meant to convey meaning, but simply to provide a support for the message.

38. Pp. 159-183.

mood,[39] to rebut statements appearing in contemporary literature
that were in contrast with basic diplomatic concepts,[40] to introduce
ideas that, although still in an embryonic state, were becoming the
focus of my research efforts,[41] to explain the ways in which diplomatic knowledge could contribute to the accomplishment of archival functions,[42] and to exhort North American colleagues to
continue the quest that I had initiated, the search for a rigorous,
systematic, body of ideas that would build upon diplomatic concepts, principles and methods, develop and adapt them, and give
origin to more ideas consistent with both the original ones and
those evolved from them.

This was the evolution of my thinking about diplomatics from
my arrival in North America to the publication of the last article of
the series. Afterwards, my thinking on the use of diplomatics, as
well as my perspective on the series and on some of the concepts
in it described continued to shift and change, partly due to external
influences. I will try to identify such influences and assess their
effects, but I have to acknowledge at the outset that, on the one
hand, I am too close to the facts to see them clearly, and on the other,
my journey of discovery is ongoing, and it is difficult to predict the
direction that it will take.

My effort at illustrating diplomatics to my North American
students and colleagues was hardly over when its product began
to acquire a life of its own. In July 1992 and 1993, two seminars
jointly held respectively in Paris, France, at the Ecole des chartes,
and Ann Arbor, Michigan, at the Bentley Historical Library, focused on the contribution of diplomatics to contemporary archival
science. While the origin of these initiatives is quite separate from

---

39. See for example the discussion on method, form and status of transmission,
particularly as it relates to facsimiles and electronic records, pp. 164-168.

40. See for example the discussion on the concepts of transaction and communication, pp. 168-170. An example of "creative rebuttal" is provided by the statements
about virtual documents at pp. 171-172 and in note 20 on page 171.

41. For example, the idea that, "where records creation is uncontrolled, diplomatics guides the establishment of patterns, the formation of a system," is at the origin
of all my subsequent research (p. 175). This statement amounted to saying that the
major future use of diplomatics was as a standard for the design of recordkeeping
and records-preservation systems.

42. In relation to description and appraisal at pp. 177-182.

the publication of my series of articles, the series itself was used as a major point of reference for most of the papers.[43] However, as at the time I did not know about these seminars, I was quite contentedly focusing my energies on putting the six articles together with examples of documents in a kit for the MAS students, changing once again the curriculum, and...writing about something else.

Thus, when, at the September 1992 International Council of Archives (ICA) meeting in Montreal, colleagues from Israel, Spain, Norway, Ireland, and Argentina (to name just a few as representative of the variety of nationalities involved) came to me expressing their appreciation for the work I had done—some of them asking whether I would give permission to translate the series—I was completely flabbergasted and a little embarassed. Firstly, and undoubtedly naively, it had never occurred to me that colleagues from other continents were reading my *Archivaria* articles. I am certain that, if I had realised that I was writing for a readership broader than my Canadian students and a few interested North American colleagues, I would have been so intimidated by the task that I would not have even begun it. Secondly, I would never have suspected that my European colleagues, surely already knowledgeable about classic diplomatics, in the absence of the kind of challenge that had stirred my efforts, would be interested in reading my articles, let alone find them useful for their own purposes. Thirdly, I would not have thought that European concepts adapted to a modern North American environment could be relevant to archivists from other traditions. Fourthly, I did not believe that my creative efforts were as "creative" as my international colleagues seemed to believe, and the basis for the establishment of a "contemporary diplomatics" quite distinct from the classic one. Fifthly and finally, I realised with both surprise and dismay that my European

---

43. Bruno Delmas and Francis Blouin state so in the introduction to the proceedings of the two seminars, published in French in *La gazette des archives* and forthcoming in English in *The American Archivist* 59 (Fall 1996). See "De la diplomatique medievale a la diplomatique contemporaine. Actes du colloque organise par l'Ecole nationale des chartes et la Bentley historical Library de l'universite de Ann-Arbor (Michigan, Etat-Unis). Paris, 6-10 Juillet 1992 et Ann-Arbor, 5-9 juillet 1993. Publies avec le concours de l'Ecole nationale des chartes." In *La gazette des archives* 172 (Ier trimestre 1996): 7-106.

archival colleagues were regarding my first three articles, those written only for purposes of explanation of the core diplomatic concepts, as a major reinterpretation of traditional diplomatics. Even if some diplomatists I had the opportunity to meet on other occasions did not completely agree with my archival colleagues as to the degree of innovation I had introduced into the concepts presented in the initial articles, I now admit and consider it only natural that, attempting to make those concepts understandable to North Americans, I did in some measure interpret them for my readers, present them in accessible terms, and emphasise the components that were more applicable to their reality and more familiar.

As a consequence of this discovery, whenever invited as a speaker outside North America, I chose to discuss the idea of a "contemporary diplomatics" and the possible ways of developing such a body of knowledge, especially in relation to electronic records.[44] In the same period, it occurred to me that my intuition that diplomatic concepts and principles could be used as a set of international standards needed to be tested.[45] This was the time of the *Armstrong v. Executive Office of the President* case and of the

---

44. Some of the presentations I made in 1993 on this subject resulted in 1994 publications. See for example: "La definizione di memoria elettronica: il passo fondamentale nella sua preservazione," *L'eclissi delle memorie*, Tullio Gregory e Marcello Morelli eds. (Bari: Editori Laterza, 1994), 147-160; "Registros Documentais Contemporaneos como Provas de Acao," *Estudos Historicos. 20 CPDOC anos*, Fundacao Getulio Vargas ed. (Rio de Janeiro: Fundacao Getulio Vargas, 1994), 49-64; "Caratteristiche intrinsiche degli strumenti informatici," *Rassegna degli Archivi di Stato* a. LIV, 1 (Roma: Ministero per i Beni Culturali e Ambientali, 1994): 57-65.

45. Standards are sets of rules of co-operation between and among peer entities that are independent of any particular context. Anything capable of being reduced to rules, that is, to principles to which actions will conform, is capable of being standardised. If archival documents were capable of being reduced to rules, to a set of characteristics or attributes, they could be created, captured, handled and managed in accordance with international standards. Diplomatics as a science is rooted in the assumption that all archival documents can be analysed, understood and assessed in terms of a system of formal elements that are universal in their application and decontextualised in nature. The primary contribution of diplomatics to modern record making and recordkeeping could be its definition of the archival document, or record, in its own terms, by rules or attributes that have evolved out of a scientific study of documentary process. Moreover, because, in the course of identifying those attributes, diplomatics has provided definitions of concepts nec-

decision of American Judge Charles Richey that some materials created by and/or stored in electronic information systems are to be considered records and treated as such. That decision merely recognised the increasing difficulty of determining what could serve as evidence of action and decision in both public administration and private business. To my mind, the solution to that problem could not derive from purely pragmatic and ad hoc decisions, but needed to be rooted in principles and criteria that could be applied in different situations and various contexts. What better opportunity could there be for testing the capacity of diplomatic concepts and principles to serve as international standards for identifying which of the materials created, transmitted, and preserved by electronic information systems qualify as records, and for designing recordkeeping and record-preservation systems ensuring the creation, handling and maintenance of reliable and authentic records?

I decided to discuss the possibility of a research grant application with my colleague Terry Eastwood. He expressed his belief that, for the purposes of the research project I had in mind, we could not rely only on diplomatics, but we needed also archival science. After all, archival science had emerged out of diplomatics as a consequence of the archival effort to address documentary and functional relationships between and among records aggregations and to study the ways in which these are controlled and communicated. We agreed that, in the course of the research, the principles and concepts of diplomatics had to be integrated with those of archival science and interpreted within the framework of electronic systems. We submitted a research proposal reflecting this decision, obtained funding, and spent the following three years carrying out

---

essary to guide the implementation of systems for creating, maintaining and preserving electronic records, it could become a design tool. These conclusions, which I drew explicitly in the *Sixth Article*, but underlay the entire series—as revealed by the emphasis put on records creation and records components—were only based on logic reasoning and were not supported by experiential knowledge.

a research project entitled "The Preservation of the integrity of electronic records."[46]

The conceptual analysis of electronic records and the project's findings have confirmed that diplomatics provides a powerful and internally consistent methodology for preserving the integrity of electronic records and is capable of constituting a reliable international standard for the design of records systems controlling electronic and non-electronic records in an integrated way. Moreover, the fact that, in the course of the research, the conceptual analysis was situated within a knowledge engineering framework has resulted in a fruitful re-examination and adaptation of diplomatic ideas in light of the electronic records reality, and in the building of a body of knowledge that can be considered the nucleus of "contemporary diplomatics."

This "contemporary diplomatics" though does not have the purity of classic diplomatics as it results from the theoretical and methodological integration of diplomatics and archival science. And quite naturally so, as, differently from the original diplomatists, who were dealing with isolated documents, modern diplomatists are concerned with aggregations of documents. Thus, the science of the records has gone full circle: archival science, which grew out of diplomatics in the nineteenth century, has reabsorbed

46. This research project, carried out between April 1994 and March 1997, was funded by the Social Sciences and Humanities Research Council of Canada (SSHRCC). For information about the hypotheses, methodology, and findings of the project, see Luciana Duranti and Terry Eastwood, "Protecting Electronic Evidence: A Progress Report," *Archivi & Computer* 5:3 (1995); Luciana Duranti, Heather MacNeil and William E. Underwood, "Protecting Electronic Evidence: A Second Progress Report on a Research Study and Its Methodology," *Archivi & Computer* 6:1 (1996): 37-70; Kenneth Thibodeau and Daryll R. Prescott, "Reengineering Records Management: The U.S. Department of Defense, Records Management Task Force, *Ibid.*, 71-78; Heather MacNeil, "Implications of the UBC Research Results for Archival Description in General and the Rules for Archival Description in Particular," *Archivi & Computer* 6:3-4 (1996): 239-46; Luciana Duranti and Heather MacNeil, "Protecting Electronic Evidence: A Third Progress Report on a Research Study and Its Methodology," *Archivi & Computer* 6:5 (1996): forthcoming; Heather MacNeil, "Protecting Electronic Evidence: A Fourth Progress Report on a Research Study and Its Methodology," *Archivi & Computer* 7:2 (1997): in press; Luciana Duranti and Heather MacNeil, "The Protection of the Integrity of Electronic Records: An Overview of the UBC-MAS Research Project," *Archivaria* 42 (Fall 1996): in press. Other writings on the research findings are forthcoming.

it within its own system of ideas.[47] Long before reaping the results of the research project, this natural flowing of diplomatics of modern records into archival science was perceived as necessary in the context of the curriculum of archival education. In 1993-94, the MAS curriculum was revised again to reflect the maturation of our thinking.[48]

We realised that, in light of our present understanding, the teaching of traditional archival functions to be performed on inactive records had to follow, rather than precede, the teaching of the context of records creation and of records creation itself. Thus, it did not make any sense to teach subjects like diplomatics, records management and the juridical system in the second year of instruction, as they constituted the theoretical basis for understanding precisely that records' nature that archivists were supposed to respect and protect in the course of appraisal, arrangement and description, etc. Moreover, it appeared clear at this point that diplomatic concepts and principles had to be dealt with in the context of all courses and there was no justification for a whole course on diplomatics. Therefore, we turned around the content of the curriculum by moving anything related to records creation, handling, and use to the first year of instruction,[49] and we embedded diplomatics primarily into a first year course exploring "the nature of archival material," which also included the concepts related to records aggregations. Other courses too incorporated

---

47. For a detailed discussion of the origin and evolution of diplomatic and archival knowledge, see Luciana Duranti, "Archival Science," *Encyclopedia of Library and Information Science*, Allen Kent ed., vol. 59, supplement 22 (New York: Marcel Dekker, 1997), 1-19.

48. The "we" I am referring to includes Terry Eastwood and myself. The fact is that, within all the work we were doing together on the research proposal and on the curriculum, it was by now impossible to distinguish our respective inputs. Our thinking was very much evolving in the same direction, and from this time on, even when strictly discussing diplomatics, I cannot talk anymore in first person.

49. This is not entirely correct, because, in the second year of instruction, two courses are offered on "recordkeeping" and "archival concepts." However, these courses are historical in nature and presuppose the knowledge of the theory and methods governing records creation, handling and use.

diplomatic concepts, such as the one on "the juridical context of Canadian archives."[50]

The experience of teaching diplomatics in the context of archival science, and of conducting deductive research which has integrated diplomatic and archival concepts and methods has been instrumental in making the decision of building upon the findings of the research project nearing conclusion by designing a new research project whose questions and methodology reveal the unification of the two disciplines.[51] Also, the fact that the findings of the previous research are not dependent on context, being derived from concepts used as international standards, has allowed for the conception of a research project that is international in nature. In a sense, one could say that another important contribution of diplomatics to the modern world of records is its fostering of international cooperation at a degree not encountered before.

Has the impact of the diplomatic series gone beyond the MAS curriculum at UBC, the theses of MAS graduates, research findings yet to be fully articulated, international debate, the expression of interest of computer and information scientist, and the archival community's generic statements that diplomatic concepts will lead the way of the next generation of archivists? At this stage, I cannot say. Certainly the individuals who studied the articles as a course requirement have acquired an outlook formed by them, and I am sure that such outlook will guide their approach to archival work for many years to come. That component of the international archival community which is knowledgeable in classic diplomatics and regularly uses its concepts in the course of its work appears eager to test the ideas expressed in the articles. Archivists who have never attended diplomatic classes and have never been formally educated in diplomatics seem to have absorbed in varying degrees diplomatic terms, type of analysis, concepts, principles, or perspec-

50. For a description of the new curriculum see Terry Eastwood, "Revised MAS Curriculum at UBC," *ACA Bulletin* 18, 5 (May 1994): 14-16. See also, by the same author, "Reforming the Archival Curriculum To Meet Contemporary Needs," *Archivaria* 42 (Fall 1996): in press.

51. This new project, entitled "The Long-Term Protection of the Reliability and Authenticity of Inactive Electronic Records," if funded, will begin in January 1998, and will be carried out by an international group of researchers.

tive and attempt to use them for their own purposes, and even to be creative and build upon them.[52]

Thus, the next question becomes: would I write the diplomatic series again? When I was asked permission to publish the articles as a collection in Spanish translation destined first to South America and then to Spain, I did not reflect on the issue, probably because this book of diplomatics would join several other books of diplomatics in the library of Spanish speaking archivists.[53] However, when I received the same request for a volume destined to the North American audience, I hesitated. I was concerned about the fact that my series would be the only such text available in English, and therefore the practically exclusive source of diplomatic knowledge for the North American archival community. To my mind, the main problem was that the articles are frozen in their temporal context: they each reflect my thinking at a particular point in time, as I have tried to show. How were readers going to know that my thinking has matured since the writing of the last article? I was told that I could exercise the option of editing the articles. I thought about it, and concluded that, if given the opportunity to touch their text, I would not stop before rewriting them completely. But what would the purpose of such an undertaking be? If I was going to the

---

52. It is interesting that the Canadian person who, in my judgment, has best penetrated the depth of diplomatics, and acquired the capacity of using its rigour for the most varied aims and even of developing its concepts to their ultimate consequences, has never been a student in my diplomatics classes and has never received a formal education in diplomatics: Heather MacNeil is now expressing her understanding of diplomatic concepts in her doctoral dissertation on "Documentary Evidence as the Verification of Truth by means of Proof," but has already shown in several writings, specifically those discussing descriptive standards, her command of the diplomatic discipline, based mostly on the series of articles. It is also very rewarding for me to open the pages of archival journals and find articles using the diplomatic ideas expressed in my series in a perfectly correct and constructive way [see for example Tom Belton, "By Whose Warrant? Analysing Documentary Form and Procedure," *Archivaria* 41 (Spring 1996): 206-220], or attend a Society of American Archivists conference and discover an entire session on the use of diplomatic analysis.

53. The Spanish volume is: Luciana Duranti, *Diplomatica. Usos nuevos para una antigua ciencia*. 1a. edicion en castellano. Traduccion, prologo y presentacion de Manuel Vazquez (Carmona, Sevilla: S&C ediciones, 1996). This same translation had been previously distributed to the Argentinian and Peruvian archival communities (Cordova, 1995).

trouble of reformulating the articles, I might as well write a new book. Therefore, the answer to the question of whether I would write again the diplomatic series is negative: today, I would write a book of contemporary diplomatics instead. However, the issue of what to do with the existing articles remained. Once again, the answer came from my students: "just tell the story in an introduction, and leave the articles as they are!"

The final question is: if my thinking on diplomatics and its significance for the management of modern records has so dramatically matured since I wrote the articles, why should anyone bother reading them, rather than wait for the new book, reflecting the present state of "contemporary diplomatics"? The answer has something to do with the fact that diplomatics is an old science and any new use of it would be impossible without a thorough understanding of the original concepts and their application. It also would have something to do with the formative function of the study of classic diplomatics. But, mostly, it concerns the validity and endurance of any present and future effort of adaptation and development of diplomatic concepts, mine as well as others'. This effort can only be evaluated and its results either used or dismissed if archivists have a solid grasp of diplomatic ideas and of the way they have been interpreted, made explicit, adjusted, expanded, and added to. In other words, the capacity to read critically any new proposal is firmly rooted in the understanding of its roots.

Thus, dear readers, here they are, my six articles of diplomatics, and, with them, the story of my long journey of discovery, not just of diplomatics, but of my new archival world. I had come to North America as an enthusiastic educator ready to leave behind heavy and tired traditions and old sciences only to find that I could not understand this modern archival world other than by turning again to those traditions and sciences and looking at them through the eyes of both an old European and a new North American.

# Chapter 1

# The Origin, Nature and Purpose of Diplomatics

Diplomatics is the study of the *Wesen* [being] and *Werden* [becoming] of documentation, the analysis of genesis, inner constitution and trans-mission of documents, and of their relationship with the facts represented in them and with their creators. Thus, it has for the archivist, beyond an unquestionable practical and technical value, a fundamental formative value, and constitutes a vital prelude to his specific discipline, archival science.[1]

This is the first of a series of six chapters which examines diplomatic doctrine from the point of view of the contemporary archivist. The whole work is directed to those who have little familiarity with diplomatics, and is meant to give them the basis for a fruitful consultation of specialized literature. However, it is also directed to those who have known diplomatics in the context of medieval studies and appreciate its potential for the identification, evaluation, control, and communication of archival documents.

This first chapter defines the science of diplomatics, looks at its origin and historical development, explores its character as it relates to documents, and, while discussing its purposes, analyzes the concepts of authenticity and originality. The five following chapters will concentrate on 1) the concepts of fact and act, and the

---

1. Giorgio Cencetti, "La Preparazione dell'Archivista," in *Antologia di Scritti Archivistici*, ed. Romualdo Giuffrida (Roma: Ministero per i beni culturali e ambientali. Pubblicazioni degli Archivi di Stato, 1985), p. 285. All translations from the Italian are by the present writer.

function of a document in relation to facts and acts; 2) the persons concurring in the formation of a document, and its nature in relation to them; 3) the genesis of public and private documents; 4) the intrinsic and extrinsic elements of documentary forms; 5) the methodology of diplomatic criticism, and the use of diplomatic analysis for carrying out individual archival functions.

The approach will be fundamentally theoretical, although an effort will be made to illustrate concepts with examples and to make connections with realities well known to North American archivists. The question which will be present all along in the mind of the readers, "How am I to use all this?," will probably be indirectly answered as the exposition of doctrine proceeds; in any event, it will be directly addressed in the last chapter.

## Why This Book?

"The most vital question" for contemporary archivists is what constitutes the body of knowledge that belongs to and identifies their profession.[2] While the education of European archivists, although incorporating historical, administrative, and legal elements, is founded on diplomatics and paleography, North American archivists have grounded their work essentially on the knowledge of history and the history of administration.[3] Nevertheless, often without fully realizing it, in a natural way, the latter have paid attention to the object of diplomatics and paleography, namely the forms and script of documents, even if unsystematically and inconsistently, more feeling their way than seeing it. This happened not only because an archives is a whole constituted of parts and it is impossible to understand and control the whole without understanding and controlling its parts, even the most elemental of them, but also because of the historical knowledge of North American archivists. In fact, history, and particularly the

---

2. Terry Eastwood, "Nurturing Archival Education in the University," *The American Archivist* 51 (Summer 1988), p. 229.

3. Luciana Duranti, "Education and the Role of the Archivist in Italy," ibid., pp. 346-355. This issue of *The American Archivist is* entirely dedicated to archival education, and shows the basic differences in approach between Europe and North America.

history of administration and law, like paleography and archival science, derived as scientific disciplines which use primary sources from diplomatics, and, in the process of becoming autonomous sciences in their own right, used principles and methodologies of diplomatics and paleography and adapted them to their own purposes, incorporating them into their own methods. As a consequence of these developments, diplomatics as an independent science came to restrict its area of enquiry to the chronological limits of the medieval period, joining paleography which was confined within those same limits by the object of its study.

However, the principles, concepts, and methods of diplomatics are universally valid and can bring system and objectivity to archival research into documentary forms, that is, a higher scientific quality. It is well known that the archivist's research into the nature or character of records has purposes different from that of the historian. Thus it is not advisable for archivists to adopt diplomatic methodology as it has been filtered through the needs of scholars of history. Rather, it is appropriate for them to extract directly from the original science of diplomatics those elements and insights which can be used for their work, and to develop them to meet contemporary needs.

It was in the 1960s that diplomatics and archival science were divorced from an exclusive association with historical sciences. A jurist, Massimo Severo Giannini, in his lectures on administrative law, as recalled by Leopoldo Sandri in 1967, used to teach that "among the non-legal disciplines which study administrative facts, there are some which analyze these facts specifically, because they [the facts] have properties that no other science or discipline has the function or instruments to analyze. The most ancient of these disciplines are accounting, archival science, and diplomatics." As Sandri himself then put it: "thus, the other face of the moon, that is archival science as the discipline which studies specific facts related to administrative activity, imposes itself on our attention, and the combination, from this point of view, of archival science and diplomatics is not less important to us."[4] Indeed, it is even more important twenty years after that remark. Nevertheless, the use of

---

4. Leopoldo Sandri, "L'Archivistica," in *Antologia di Scritti Archivistici*, p. 21.

diplomatics by contemporary archivists will no doubt face serious difficulties.

It has often been pointed out that it is extremely difficult to comprehend recent events. Part of the reason is undoubtedly that our society creates sources of information which emerge in forms at the same time manifold and fragmentary. We are engulfed and bewildered by it all. Moreover, as Italian archivist Paola Carucci has noticed, even when it is possible to posit lines of development or critical phases in the documentation function, it is often impossible to verify them for our own time because we lack the proper perspective on events in which we are still involved, and, given the multiplicity and variety of information, the knowledge of a single document is rarely determinant. It is often necessary to assemble a panoply of different sources of information in order to understand any given document, each of which, by itself, may appear of scant utility.[5]

A major problem is created by the fact that the number of actions and events taking place exclusively in a personal sphere is limited. From birth to death, written traces of persons can be found anywhere. A great many bodies produce documentation about the same person or event. This phenomenon has an important impact on the process of identification and selection of the sources which it is appropriate to preserve. The choice is conditioned by the culture and the historical-legal-administrative sensibility of the archivist, but also and foremost by the ways current records are formed and maintained.

Thus, if the knowledge of administrative structures, bureaucratic procedures, documentary processes and forms (that is, of administrative history, law and diplomatics) allows archivists to make a comparative analysis of archival series for selection and acquisition, that same knowledge enables them to participate with competence in the creation, maintenance, and use of current records by giving advice about the determination of document profiles, the simplification of bureaucratic procedures, and the adoption of classification and retrieval systems.

---

5. Paola Carucci, Il *Documento Contemporaneo. Diplomatica e Criteri di Edizione.* (Roma: La Nuova Italia Scientifica, 1987), p. 11.

However, the use of diplomatic criticism for records management and appraisal functions particularly requires a development of *special diplomatics*, and here lies the major difficulty that diplomatics encounters in its evolution as a discipline for contemporary records.

Special diplomatics is a branch of diplomatics, a discipline in which "the theoretical principles formulated and analyzed by diplomatics individualize, develop and clarify themselves being applied to single, concrete, real, existent and easily exemplifiable documents, rather than to an abstract and atypical general documentation." In Georges Tessier's words: "â côté d'une diplomatique générale ayant pour objet les notions fondamentales et l'exposé de la methode, on peut concevoir autant de diplomatiques spéciales que de foyers ou de courants de civilisation."[6] Thus, general diplomatics is a body of concepts. The application of them to infinite individual cases constitutes the function of diplomatic criticism, that is, of special diplomatics. Theory (general diplomatics) and criticism (special diplomatics) influence each other. The latter, analysing specific situations, uses the former; the former guides and controls and is nourished by the latter.

The body of principles and methods as established in the nineteenth century manuals of diplomatics does not need to be reformulated for the criticism of contemporary documents, but merely re-examined and adapted. However, the development of special diplomatics for contemporary documents cannot derive simply from the direct application of that theory to single documents, because of all the problems presented by the plurality and fragmentation of our sources, and because the formalism of old bureaucracies has atrophied in modern ones, creating forms of documents which do not often lend themselves to systematic analysis and description.

It is not accidental that archivists' interest in diplomatics has occurred at the moment of maximum development of records management. This new discipline is very old indeed, as witnessed

---

6. Cencetti, "La Preparazione dell 'Archivista," 286; Georges Tessier, "Diplomatique," in *L'Histoire et ses méthodes*, ed. Charles Samaran (Paris: Librairie Gallimard, 1961), p. 668.

by the series of little treatises entitled *De Archivis* that appeared during the seventeenth and eighteenth centuries, all full of advice on the creation, arrangement, and description of *current* documents. And we have to remember that formularii and regulations of chanceries had already been in existence for centuries. Only at the time of the French Revolution did archivists move from the management of current records to the care of "historical sources", to which they tried to apply the classification principles learned in administrative offices.[7]

Hence, the development of the two disciplines of records management and diplomatics is inextricably linked. When there are rules governing the genesis, forms, routing, and classification of documents, special diplomatics can identify the rules through the criticism of documents. On the basis of those rules, it can establish the value of the examined documents. Thus, the expansion of records management feeds special diplomatics. However, the opposite is also true. Where there are not records management rules in place, the study of diplomatic principles and methods gives to those who try to formulate them a clear indication of the elements which are significant and must be developed, while the examination of various special diplomatics of past administrations in different societies gives them the critical judgement deriving from comparative study.

The reciprocal influence of records management and diplomatics can indicate the road to take in the future, but how can we develop a special diplomatics for the documentation created in the period between the French Revolution and our days, part of which documentation we still have to appraise, arrange, and describe? The application of diplomatic criticism to the records of the last two centuries requires a specific study of the records-keeping practices of each single administration, which is more than and somehow different from administrative history.

The study, through the examination of laws, regulations, and archival documents, of the way records creators organized their

---

7. Luciana Duranti, "The Importance of Records Managers to Society," *Vanarma*, 18 (March and April 1988); "The Odyssey of Records Managers," *Records Management Quarterly* (1989), in press.

memory is in Italy the specific function of a discipline called "special archival science," being the application of archival theory to individual cases. Between archival science and special archival science there is the same relationship that links diplomatics and special diplomatics. Thus, archival science is the doctrine, while special archival science is the criticism, which, directed and controlled by the doctrine, represents the reaction of scientific minds coming into contact with series and fonds. Moreover, special archival science, compared to administrative history, has a profound juridical nature, being largely the history of the law and of its application in administrative activities, based on the analysis of the product of those activities, the archives. Where, then, is the difference between special archival science and special diplomatics? The boundary line between the two disciplines is to be found in the series, the fonds, the archives as a *complex* of documents, as a whole, which constitutes the area of archival science. Instead, the single document, the elemental archival unit, is the area of diplomatics.

The historical-administrative-legal-archival study conducted on the creators of documents is thus essential to the development of a special diplomatics of the documents of past societies. However, it is not less important for the diplomatic criticism of documents of present and future societies. Actually, its relevance is enhanced by the proliferation of laws and of administrative bodies and by the continuous change of structures and functions. But such study is easier to carry out because of the growing uniformity of those laws, regulations, structures, and of the ways activities are carried out, because of the standardization promoted by records management, which is vital to an elephantine bureaucracy, and because freedom of information, underlining the accountability of administrative bodies and the citizens' right to control their activities, favours a better organization and determines the spreading of the knowledge of our social system, knowledge which is losing its elitist character.

To say that special archival science or, if you want, history of administration and its documentation and history of the law, constitutes the necessary mediation between diplomatic theory and its application to concrete, real cases does not mean that a full development of those studies must precede the exercise of diplomatic criticism of the documents of a specific body or person. Rather, it means that whoever undertakes such an analysis in order to understand diplomatically those documents needs to investigate the meaning of their forms not only in the individual context of the creator but in the broader context constituted by the legal doctrine of the creator's society and its manifestation in the documentation function of that society. For instance, only in that broader context is it possible to grasp the real meaning of documents' names and so their nature. Consider that, today, types of documents are defined sometimes in relation to the legal nature of the action giving origin to them (sentences, permits, concessions, and contracts) and sometimes in relation to their form (letters, minutes, notes, and indentures). The study of the relationship between the nature of the action generating a document and the form of the document is one specific object of diplomatics, but it is only a tile in a very complex mosaic, which has to be reconstructed with the assistance of allied disciplines.

The history of administration and of its documentation function and the history of the law and of its manifestations, which I call special archival science, archival science, and general and special diplomatics together constitute a constellation of disciplines each of which increases the light provided by the others. The use of diplomatic criticism can give a substantial and unique contribution to the management of current and semicurrent records and to the identification, appraisal, arrangement, description, and communication of inactive ones, both public and private, but such a contribution would not be possible without the auxilium of those other disciplines which constitute the body of proper knowledge of the archivist. Thus, when an archivist studies records, whatever discipline he decides to use or whatever the specific object he chooses to investigate, his study will always have an historical-administrative-legal-diplomatic character, that is, his study will have an archival nature.

Many North American archivists are conscious of all this and have called for research into the subject.[8] This book attempts to

<hr>

8. Hugh Taylor, "Transformation in the Archives: Technological Adjustment or Paradigm Shift?", *Archivaria* 25 (Winter 1987-88), p. 18; Barbara Craig, "Meeting the Future by Returning to the Past: A Commentary on Hugh Taylor's Transformations," *ibid.*, p. 9. In this regard, an interesting article is David Bearman and Peter Sigmond, "Explorations of Form of Material Authority Files by Dutch Archivists," *The American Archivist* 50 (Spring 1987), pp. 249-253.

answer that call. However, the virtual non-existence of literature on diplomatic theory as it applies to modern and contemporary documents makes of this re-examination and adaptation of principles, concepts, and methods little more than a tentative exploration of new ground, aimed mainly to stimulate reactions, further thinking, and research.

## The Word Diplomatics

In many libraries, diplomatic literature is incorrectly classified under either diplomacy or paleography, *not* under diplomatics. There are etymological-historical reasons for the confusion of diplomatics with diplomacy, and there are scientific reasons for its confusion with paleography.

Both the words diplomacy and diplomatics have their root in the Greek verb *diploo* ( διπλόω ), meaning 'I double' or 'I fold', which gave birth to the word *diploma* (διπλόωμα), meaning 'doubled' or 'folded.' In classic antiquity, the word diploma referred to documents written on two tablets attached with a hinge and called *diptych;* and, during the Roman Imperial period, to specific types of documents issued by the Emperor or the Senate, such as the decrees conferring privileges of citizenship and marriage on soldiers who had served their time. In time, diploma came to mean a deed issued by a sovereign authority, and was extended to include generally all documents issued in solemn form.

The term diplomatics is a modern adaptation of the Latin *res diplomatica*, the expression used by the first writer on the subject to refer to the critical analysis of the forms of diplomas. The term diplomacy, from the French *diplomatie,* refers to the art of conducting international negotiations, which result in the compilation and exchange of official documents, namely diplomas.

The confusion between diplomatics and paleography is not of a terminological nature, but is deeply rooted in the history of the two disciplines and in the philosophical conceptions of the eighteenth century.

## The Origin and Development of the Discipline

Diplomatics and paleography were born as sciences arising from the need to analyze critically documents considered to be forgeries.[9] The problem of distinguishing genuine documents from forgeries was present in the earliest periods of documentation, but until the sixth century no attempt was made to devise criteria for the identification of forgeries. Even legislators did not demonstrate interest in the issue, basically because of the legal principle commonly accepted in the ancient world that authenticity is not an intrinsic character of documents but is accorded to them by the fact of their preservation in a designated place, a temple, public office, treasury, or archives. This principle was open to abuse. Eventually, people began to present forgeries to designated records offices to lend them authenticity. Therefore, practical rules to recognize them were introduced in Justinian's civil code *(Corpus iuris civilis)* and later in a number of Papal *Decretales.* These rules referred only to the external forms of documents created by imperial and papal chanceries, that is, to documents contemporary to the laws, not documents of previous centuries which were often used by authorities to support political or religious claims.

In time, largely as a result of controversies over the authenticity of these very political or religious claims, humanist scholars began to apply to documentary texts a sophisticated criticism based on historical methodology. Using this type of systematic analysis, the Italian Renaissance humanists Francesco Petrarca and Lorenzo Valla in the fourteenth and fifteenth centuries respectively, proved that the privileges granted to Austria by Caesar Augustus and Nero in the first century and the donation made by Constantine to Pope Silvester in the fourth century were forgeries.

The transformation of the critical analysis of the document into a complete and autonomous discipline was determined by the

9. For the history of diplomatics I have relied in particular on Tessier, "Diplomatique," pp. 633- 676; Alessandro Pratesi, *Elementi di Diplomatica Generate* (Bari: Adriatica Editrice, n.d.), pp. 9-19; Giulio Battelli, *Lezioni di Paleograf a* (Città del Vaticano: Pont. Scuola Vaticana di Paleografia e Diplomatica, 1949), pp. 11-24.; Francesco Calasso, *Medio Evo dew Diritto. I—Le Fonti* (Milan:Giuffre,1954), pp. 57-95 and pp. 301-408.

so-called 'diplomatic wars' (*Bella diplomatic*), which took place in the seventeenth century and concentrated attention on specific groups of documents. In Germany, the diplomatic wars were judicial controversies over the affirmation of a right, while in France they assumed a doctrinal character with a serious scientific concern: they prepared the ground for the great debate between the Benedictines of the Congregation of Saint-Maur in France and the scientific society founded in Antwerp by the Jesuit Jean Bolland.

In 1643, the Bollandists began to publish the first volumes of a colossal work, the *Acta Sanctorum*, in which the testimonies related to the lives of single saints were evaluated for the purpose of separating the facts from the legend. Its second tome appeared in 1675 with an introduction written by Daniel Van Papenbroeck, in which the general principles for establishing the authenticity of old parchments were rigorously enunciated. However, applying those principles to the diplomas of the Frankish kings, Papenbroeck erroneously declared a diploma of Dagobert I to be a forgery and in doing so brought into discredit all the Merovingian diplomas, most of which were preserved in the Benedictine Monastery of Saint-Denis. Dom Jean Mabillon, Benedictine of the Congregation of Saint-Maur, who had been called from the Monastery of Saint-Denis to the Abbey of Saint-Germain-des-Pres to publish the lives of Benedictine saints, answered the accusation of Papenbroeck six years later, in 1681, in a six-part treatise, *De Re Diplomatic Libri VI*, which established the fundamental rules of textual criticism.[10]

The publication of Mabillon's work marks the birth date of diplomatics and paleography. Mabillon subdivided a group of about two hundred documents into broad categories and examined all the different aspects which could be analysed: material, ink, language, script, punctuation, abbreviations, formulas, subscrip-

---

10. Daniel Van Papenbroeck, *Acta Sanctorum Aprilis* (Antwerp, 1675). The *Acta Sanctorum Quotquot Toto Orbe Coluntur* presently includes 67 volumes, the first 50 published in Antwerp, I in Tongerloo, and the others in Brussels. It is still being added to. Two other publications of this work have been initiated in Venice and in Paris. Dam Jean Mabillon, *De Re Diplomatica libri VI...* (Paris, 1681); *Librorum de Re Diplomatica Supplementum* (Paris, 1704); *De Re Diplomatica libri Al...*, editio secunda ab ipso auctore recognita, emendata et aucta (Paris, 1709); *De Re Diplomatica Libri VI...*, tertia atque nova editio ...,2 vols., (Naples, 1789).

tions, seals, special signs, chancery notes, and so on. If five parts of the treatise contain mainly diplomatic criticism, one entire part is dedicated to the analysis of the script and can be considered the first treatise on paleography. However, the science which studies ancient scripts did not yet have a name; the term paleography was coined by another Benedictine, Dom Bernardo de Monffauçon, who published *Palaeographia graeca, sive de ortu et progressu literarum* in 1708, but the systematic study of types of script was initiated by Mabillon.

If the impetus for articulation of a method of proving the authenticity of documents came from doctrinal conflicts of the Reformation and Counter-Reformation, that is, from a practical need, the development of the discipline so created soon rose above the religious fray. As long as documents were considered exclusively as legal weapons for political and religious controversies or in disputes before the courts, the methodology of textual criticism was utilitarian in nature, and was therefore looked upon as being suspect; but when scholars began to look at documents as historical evidence, diplomatics and paleography acquired a scientific and objective character. However, given the encyclopedic conception of knowledge that dominated the eighteenth century, they remained confused as one discipline for a long time.

In the middle of the century, the teaching of diplomatics and consequently of paleography was introduced in university faculties of law, and this led to the publication of numerous works on the subject in Germany, France, England, Spain, and Italy. The works conceived in academic schools tended to present an excess of schematization that reached its apex in the attempt of Johann Christoph Gatterer, professor at the University of Gottingen, to introduce to diplomatics a version of the classification system adopted by Linnaeus in the natural sciences.[11]

---

11. Johan Christoph Gatterer, *Elementa artis diplomaticae universalis* (Göttingen, 1765). Between Mabillon's and Gatterer's, other works of some importance are: Dom Giovanni Perez, *Disser- tationes ecclesiasticae de re diplomatica* (n.p. [Spain], 1688); Thomas Madox, *Formulare Anglicanum* (London, 1702); Scipione Maffei, *Istoria diplomatica che serve d'introduzione all' arte critica in tal materia* (Mantua, 1727); Dom Johan Georg Bessel, *Chronicon Gotwicense* (Gottweig, 1932), which, examining the characteristics of Imperial and Royal German documents, offered the first example of *special diplomatics*.

Notwithstanding the fervour of study in the universities, once again the greatest progress was made by two Benedictine fathers of the Congregation of Saint-Maur, Rene Prosper Tassin and Charles Toustain, who published in Paris, between 1750 and 1765, the six-volume *Nouveau traité de diplomatique*. The authors investigated many documents going back to the first centuries of the Middle Ages and having their origin beyond the boundaries of France. In so doing, they also entered the field of special diplomatics. Their critical history of documentary styles, formulas, and uses, and the principles of methodology they introduced, are still valid today.

The *Nouveau traité* was translated into German during its compilation.[12] This peculiar phenomenon demonstrates not only "la solidarité internationale dans le domaine de la culture au XVIIIe siècle,"[13] but also the scientific validity of diplomatic principles and methodology for the criticism of all documents independently of time and place of creation.

The nineteenth century saw the creation of the "École des Chartes" in Paris in 1821, the consequent development of paleography into an autonomous discipline,[14] and decisive progress in the formulation and definition of diplomatic principles. However, the greatest advances took place in Germany and Austria where the flowering of historical studies was more significant than in France. In 1831, the publication by Johan Friedrich Bohmer of a complete chronological catalogue of the documents issued by the Emperors of the Holy Roman Empire, with indication of their content and of their diplomatic character, initiated a period of feverish description of medieval documents and a remarkable proliferation of studies of special diplomatics.[15]

---

12. Christoph Adelung and Adolph Rudolph, *Neues Lehrgebäude der Diplomatik*, 9 vols. (Erfurt, 1759-69).

13. Tessier, "Diplomatique," p. 645.

14. The first autonomous work of paleography after Montfaucon's was Natalis de Wailly, *Éléments de paléographie pour servir à l' étude des documents inédits sur l' histoire de France*, 2 vols. (Paris, 1838).

15. Johan Friedrich Böhmer, *Regesta chronologico-diplomatica regum atque imperatorum romanorum...* (n.p., 1831). The most important publication that followed Bohmer's was Philippe Jaffe, *Regesta pontif cum Romanorum ab condita Ecclesia ad annum post Christum natum 1198* (n.p., 1851; the second edition, amplified by W. Wattenbach, et al., was published in Berlin between 1885 and 1888).

Furthermore, the bringing together in published volumes of documents created by the same office and preserved by the various addressees opened new types of enquiries and spawned sophisticated comparative analysis. Thus, Julius Ficker, noticing inconsistencies between the date of some documents and the place where they were issued, could posit the conceptual distinction between the moment of the juridical act and the moment of its documentation; and Theodor von Sickel, comparing the documents issued by the same chancery, was able to define a rigorous method that, together with the one conceived by Ficker, based evaluation of a document on analysis of the process of its creation.[16]

The advances made by Ficker and von Sickel were an outcome of post-romantic German historicism and determined a methodology of documentary criticism and a body of principles which subsequent studies would confirm and perfect without introducing any major conceptual innovation.[17]

## The Object of Diplomatics

What, then, is diplomatics? Peter Herde writes that it is "the study of documents."[18] This definition is quite general, but has the

---

16. Julius Ficker, *Beiträge zur Urkundenlehre* 2vols. (Innsbruck, 1877-1878); Theodor von Sickel, "Beitrage zur Diplomatik" I-VIII, in *Sitzungsberichte der Raiserlichen Akademie der Wissenschaften* (Vienna, 1861-1882). In the same period, the methodology of juridical studies was for the first time applied together with that of diplomatics to the critical analysis of private documents by Heinrich Brunner, *Zur Rechgeschicte der römischen und germanischen Urkunde* (Berlin, 1880). It might be noted that modern archival science is an outgrowth of these sorts of diplomatic and juridical studies.

17. Some important manuals appeared at the turn of the century. Among them, the most relevant are Harry Bresslau, *Handbuch der Urkendenlehore für Deutschland und Italien* 2 vols. (vol. 1: Berlin, 1889; vol. 2: Leipzig, 1912-1931); Artur Giry, *Manuel de diplomatique* (Paris, 1893); Cesare Paoli, *Programma scolastico di paleografa latina e di diplomatica* (Firenze, 1888-1890). For the twentieth century, it is opportune to mention the two volumes by Alain de Boüard, *Diplomatique générale* (Paris, 1952). The absence of manuals in English is indeed noticeable. In fact, English diplomatists were much more interested in the application of diplomatic methodology to specific documentary bodies than in the development of theoretical studies. As a consequence, there is a significant literature on special diplomatics produced in England. A good bibliography of diplomatics writings between 1912 and 1971 can be found in *The New Encyclopaedia Britannica* 15th ed., s.v. "diplomatics", p. 813.

18. Ibid, p. 807.

merit of moving attention from the discipline itself to its object, the document.

What is a *document?* The term traditionally refers to a multiplicity of sources of evidence. Thus, we need to specify that diplomatics studies *the written document,* that is, evidence which is produced on a medium (paper, magnetic tape, disc, plate, etc.) by means of a writing instrument (pen, pencil, typing machine, printer, etc.) or of an apparatus for fixing data, images and/or voices. The attribute "written" is not used in diplomatics in its meaning of an act *per se* (drawn, scored, traced, or inscribed), but rather in the meaning that refers to the purpose and intellectual result of the action of writing; that is, to the expression of ideas in a form which is both objectified (documentary) and syntactic (governed by rules of arrangement).

Any written document in the diplomatic sense contains information transmitted or described by means of rules of representation, which are themselves evidence of the intent to convey information: formulas, bureaucratic or literary style, specialized language, interview technique, and so on. These rules, which we call *form,* reflect political, legal, administrative, and economic structures, culture, habits, myths, and constitute an integral part of the written document, because they formulate or condition the ideas or facts which we take to be the content of the documents. The form of a document is of course both physical and intellectual. An analogy with architecture may help clarify this vital concept. We recognize a church as such because it has a shape or physical form exhibiting certain conventional elements or features such as a bell-tower, but we identify and understand the full meaning of a particular church, its cultural context, from the way those conventional elements are expressed in its architectural design, that is, from its intellectual form. Of course, a church might not present any conventional feature and still be a church because of its content. For instance, the Sacrament might simply be on a makeshift altar in a warehouse, because public worship is forbidden. The full meaning of "church" can be captured only by reflecting on both the physical building and the arrangement of its content. Like a building, a document has an external makeup which is its physical form, an internal articulation which is its intellectual form, and a message to transmit which is its content. It is impossible to understand the message fully without understanding the makeup and articulation which the author chose to express it.

The form of a written document is, therefore, the whole of its characteristics which can be separated from the determination of the particular subjects, persons or places it is about; it is "la seule à rendre raison de la véritable nature des actes écrits."[19]

However, the object of diplomatics is not any written document it studies, but only the *archival document,* that is a document created or received by a physical or juridical person in the course of a practical activity.[20] It is true that the principles and methods of diplomatic analysis can be extended to documents expressing feelings and thoughts and created by individuals in their most private capacity. In fact social habits and routines tend to penetrate all aspects of human life, so that love letters or diaries are likely to be very similar in their physical and intellectual form to executive letters, or ship's logs. But the inner freedom of human beings is such that a strict observance of rules cannot be expected in a personal context, so that a diplomatic study of forms may reveal little about the real nature of, for instance, an amateur photograph or a mother's message. Consequently, we will explore diplomatic theory only as it applies to documents which result from a practical administrative activity, be it public or private, that is, to documents

---

19. Tessier, "Diplomatique," p. 667.

20. For the purposes of this study, the term "juridical person" is used in the sense of an entity having the capacity or the potential to act legally and constituted either by a collection or succession of physical persons or a collection of properties. Examples of juridical persons are states, agencies, corporations, associations, committees, partnerships, ethnic and religious groups, positions to which individuals are nominated, appointed or hired (the National Archivist, the Professor of Diplomatics at ..., the conservator of the Museum of ...), character groups (women, fathers, children, deceased persons), the estates of bankrupt or deceased persons, counties, and so on. In France and in Quebec, the term equivalent to juridical person is *personne morale* or *juridique.* In England, the United States, and English- speaking Canada, there is a legal distinction between "natural" and "artificial" persons which is close to the distinction between physical and juridical persons, but the jurists in those countries do not agree on a definition of the two terms. Moreover, diplomatics has developed in France, Germany, Spain, and Italy, that is, in countries where the concept of juridical, as opposed to physical, person is deep-rooted in the minds of all citizens, and diplomatic doctrine is built on it. Thus, the traditional terminology of diplomatics is maintained in this study.

archival as to the circumstances of their creation. This analysis can, of course, be used for a better understanding of documents of differing nature.

If we carefully analyze a written archival document, we discover that there is much more to it than a *medium*, a *form*, and a *content*. The circumstance of the writing implies the presence either of *a fact* and a *will* to manifest it or of a will to give origin to a fact.[21] It also indicates a *purpose*. In fact, the existence of something written, directly or potentially, determines *consequences*, that is, it can create, preserve, modify, or extinguish situations. Furthermore, the document by means of which a fact and a will determine consequences is the result of a procedure, of a process of creation, a *genetic process*, that will be reflected in the documentary form, becoming one of the constituent elements of the written archival document.

Therefore, examining a document critically, diplomatics studies the fact and will originating it as they relate to purpose and consequence, the development of its genetic process, and the character of its physical and intellectual form. The study of the content of the document is extraneous to diplomatics because it is the authenticity, validity, authority, and full meaning of the content that diplomatics strives to ascertain by looking at various elements of the document.

In a society governed in all its aspects by law (be it natural, customary, common or statutory), any fact represented in an archival document is related or referable to law, and is defined as being either juridically relevant or juridically irrelevant.[22] Diplomatics has traditionally been applied to documents which contain facts juridically relevant. Thus, Von Sickel defined the document-object of diplomatics as "the written evidence, compiled according to a determined form—that is variable depending on place, period,

---

21. In diplomatics, "fact" is not to be confused with "content", the latter being the manifestation of the former through writing. The term "content" includes the idea of representation, communication.

22. The term "juridical" is broader than the term "legal". It refers to the nature of abstract legal concepts. Thus, a "juridical transaction" is a transaction legally supposed or conceived of, to some extent irrespective of its actual existence, even if it contemplates incidents and circumstances not recognized by the law.

person, transaction—of facts having a juridical nature."[23] The same definition, with minor variations, is given by Harry Bresslau, Alain de Boüard, and Artur Giry.[24] The most precise definition of a document is provided by Cesare Paoli, and reads: "a document is the written evidence of a fact having a juridical nature, compiled in compliance with determined forms, which are meant to provide it with full faith and credit."[25]

The three fundamental requisites of the document for diplomatic study, that is, the circumstance of the writing, the juridical nature of the fact communicated, and the form of the compilation, were identified in the criticism of medieval documents. Carucci points out that they are also valid for the diplomatics of modern and contemporary documents. We can assume that Paoli's definition encompasses also preparatory or interlocutory writings, those somehow connected to the final and formal one which represents a manifestation of will aimed at a juridical consequence.[26] Some time ago, Georges Tessier suggested the same thing in his definition of diplomatics: "elle est la connaissance raisonnée des règles de forme qui s'appliquent aux actes écrits et aux documents assimilé's."[27] It is evident that Tessier wishes to broaden the area of diplomatics to all those documents which are administratively created by eliminating the juridical nature of the fact communicated from the requisites of the document for diplomatic study. In fact, we can also use the instruments provided by diplomatic theory to analyze documents containing facts juridically irrelevant as long as they are created according to a procedure, routine, or habit, and in the context of a practical activity. And at this point we are already answering the question that opened this section: what is *diplomatics?* Carucci writes: "Diplomatics is the discipline which studies the

23. Von Sickel, *Acta Regum et Imperatorum Caroiinorum* (Vienna, 1867), I, p. I .

24. Bresslau, Handbuch der Urkundenlehre, I, p. I; de Bouard, *Manuel de diplomarique, I:* 32ff; Giry, *Manuel de diplomatique, p.* 4.

25. Cesare Paoli, *Diplomatica,* 2nd ed., (Firenze: Sansoni, 1942), p. 18. Those forms are often used automatically and without awareness of their real function: thus, we autograph a typed letter often only because we consider it uncourteous to type our name.

26. Carucci,11 *Documento Contemporaneo, p.* 28.

27. Tessier, "Diplomatique," p. 667. My emphasis.

single document or, if we want, the elemental archival unit (document, but also file, registers analyzing its formal aspects in order to define its juridical nature, with regard to both its formation and its effect."[28] This definition, though accurate and appropriate, imposes on diplomatic analysis the same limits that we are trying to remove, in contrast with the statements made by the author throughout her book about the broadness of the object and multiplicity of purposes of the diplomatic criticism. Thus, the best definition of diplomatics is still the one provided by Cencetti and quoted at the beginning of this chapter, a definition which may be simplified and clarified as follows: *diplomatics is the discipline which studies the genesis, forms, and transmission of archival documents, and their relationship with the facts represented in them and with their creator, in order to identify, evaluate, and communicate their true nature.* The first part of this definition has been already illustrated. It is now necessary to analyze the second part of it, that is, to examine the purposes of diplomatic criticism.

## The Purposes of Diplomatics

The origin of diplomatics is strictly linked to the need to determine the authenticity of documents, for the ultimate purpose of ascertaining the reality of rights or truthfulness of facts represented in them.

*Diplomatic authenticity* does not coincide with *legal authenticity,* even if they both can lead to an attribution of *historical authenticity* in a judicial dispute.

Legally authentic documents are those which bear witness on their own because of the intervention, during or after their creation, of a representative of a public authority guaranteeing their genuineness.[29] Diplomatically authentic documents are those which

---

28. Carucci,11 *Documento Contemporaneo, p.* 27.

29. In law, "authentic" is defined as "duly vested with all necessary formalities and legally attested." An authentic document is called by the law "authentic act" and is defined as "an act which has been executed before a notary or public officer authorized to execute such functions, or which is testified by a public seal, or has been rendered public by the authority of a competent magistrate, or which is certified as being a copy of a public register." *Black's Law Dictionary,* Revised Ivth ed.,s.v. "authentic" and "authentic act", p. 168.

were written according to the practice of the time and place indicated in the text, and signed with the name(s) of the person(s) competent to create them. Historically authentic documents are those which attest to events that actually took place or to information that is true. The three types of authenticity are totally independent of one another. Thus, a document not attested by a public authority may be diplomatically and historically authentic, but is always legally inauthentic. A Papal brief which does not contain the expression "datum ... sub anulo piscatoris" may be legally and historically authentic, but it is diplomatically inauthentic. A certificate issued by a public authority in respect of bureaucratic rules but containing information that does not correspond to reality is legally and diplomatically authentic, but historically false. Why historically false, not inauthentic? To explain, it is first necessary to illustrate the difference between an *authentic* and a *genuine* document.

A document is "authentic" when it presents all the elements which are designed to provide it with authenticity. A document is "genuine" when it is truly what it purports to be. Thus, a sentence is legally authentic when signed by a magistrate, and it is also genuine if the signature is not counterfeit. Accordingly, a privilege which purports to have been issued by an imperial chancery is diplomatically authentic when all of its forms correspond perfectly to those prescribed by the chancery regulations, and it is also genuine if it has actually been issued by that chancery.

However, the distinction between authenticity and genuineness is not valid in a historical sense. In fact, law and diplomatics separately evaluate the forms of documents and the authors of them so that we can have an authentic document which is not genuine or vice versa.[30] In contrast, history evaluates only the

30. Traditional diplomatic theory considers diplomatic authenticity and diplomatic genuineness to be synonyms. In fact, that theory was formulated for the criticism of medieval diplomas, which had such a complex genetic process and presented such a number of formal elements introduced in them with the expressed purpose of guaranteeing genuineness that *de facto* diplomatic authenticity and genuineness coincided. However, the idea of a difference between the two, although not clearly expressed, was there, because traditional diplomatics did distinguish between a genuine document in which some forms required for authenticity were missing, and a false document presenting all of those forms, even if it ended up declaring the former authentic and the latter pseudo-original, where "pseudo" conveyed the concept of diplomatic (not yet proved historical) falsity. Nevertheless,

content of the document, so that, historically, authentic is synonymous with genuine.

Even more subtle is the distinction in the uses of the antonyms of authentic and genuine, that is *inauthentic* and *false*. The concept of inauthenticity refers to the *absence* of the requisites which provide authenticity. The concept of falsity refers to the *presence* of elements which do not correspond to reality. Those elements can be either intentionally or negligently untrue, or untrue by mistake or accident when reasonable care has been exercised.

Now, according to the argument presented above, the concept of inauthenticity can be used only in a legal or diplomatic sense, not in an historical sense. In fact, the absence of required information in the content of a document cannot compromise its historical authenticity-genuineness. Thus, a private contract that is not corroborated by a public official (the term includes notaries and lawyers) is legally inauthentic, and a letter of appointment that does not contain the conditions of appointment is diplomatically inauthentic, but a form incompletely filled out or not signed as required remains historically authentic-genuine if its content is truthful.

The concept of falsity, although valid in a legal, diplomatic, and historical sense, in each of them refers to different elements of the document. This concept is perhaps best illustrated through the example of a type of medieval forgery. In those times, documents were often destroyed by fire or lost during invasions and wars, and the rights and deeds attested in them, in the absence of any other proof, were considered non-existent. So the owners of the destroyed documents used to compile new documents containing the same information as the original ones. Any one of the documents so created is legally false because the signature and the seal are counterfeit, proving that the purported author did not sign that specific document; and diplomatically false because some formal elements imperfectly reproduce the practice of the time, or place,

---

a diplomatics which has broadened its area of enquiry to all archival documents of all times needs to specify the difference between authentic and genuine, and consequently between their opposites, because modern and contemporary documentary processes and forms are much simplified and more flexible, and the presence in modern and contemporary documents of all the forms which *usually* identify an authentic document does not give any guarantee of genuineness.

proving that the specific document was not compiled when or where it purports to have been issued; but it is historically authentic-genuine because the information the document contains is true. By analogy, a modern birth certificate accidentally bearing an incorrect date of birth is legally and diplomatically genuine, but historically false. Even if the circumstance of the historical falsity of the date of birth leads of annulment of the certificate, that does not change the fact that it was legally genuine when created.

Thus, legally and diplomatically, to say that a document is false is the equivalent of saying that it is forged, counterfeit, or somehow tampered with at some time; historically, it is the equivalent of saying that the facts described in the document are untrue.

In common language, the term authentic is often confused with the term *original*, and legal terminology favours such confusion. In fact, in law, an *authenticum is* defined as "an original instrument or writing; the original of a will or other instrument, as distinguished from a copy."[31]

Because a primary function of diplomatic criticism is to distinguish an original document from a draft and a copy for the purpose of determining the degree of authority of the document under examination, and general diplomatics describes and defines the different stages of a document's transmission,[32] it is opportune to examine the meaning of original as opposed to draft or copy, in both the legal and diplomatic sense.

English law defines an original document as "the first copy or archetype; that from which another instrument is transcribed, copied or imitated."[33] This definition could also probably apply to a first draft (it being "the first copy") or to a final draft (it being the "archetype"). In fact, the definition of *draft* as offered by the same dictionary reads: "A tentative, provisional, or preparatory writing out of any document ... for purposes of discussion and correction, which is afterwards to be copied out in its final shape."[34] In con-

---

31. *Black's Law Dictionary, s.v.* "authenticum"+ p. 168.

32. The terms "transmission" and "tradition" are used as synonyms with reference to documents to mean both their genetic process and the ways they are handed down to future generations, that is, their status.

33. *Black's Law Dictionary, s.v.* "original", p. 1251.

34. Ibid., s.v. "draft", p.582.

trast, French jurists consider an original to be an "écrit constatant un acte juridique et revêtu de la signature de la ou des parties ou de leur representant, par opposition à la copie qui en est la reproduction."[35] The French definition is as restrictive as the English one is general. In fact, most ancient documents and many modern informal documents (for example, interdepartmental memoranda) are not signed by their authors.

Diplomatics examines the concept of originality and points out the common denominators of all originals, independently of time and place of creation. The first element of originality is that indicated by the English legal definition, which derives from its etymology: the latin word *originalis* means 'primitive', first in order. The second necessary element is an element of perfection. To be original a document must be *perfect*, a term which both legally and diplomatically means complete, finished, without defect, and enforceable. A perfect document is a document that is able to produce the consequences wanted by its author, and perfection is conferred on a document by its form. With regard to its essential elements, an original is defined by Tessier as "l'exemplaire à la fois originel et parfait d'un acte quelconque."[36] We could also say that *an original is a perfect document and the first to be issued in that particular form by its creator.*

Of course, there may be more than one original of the same document created either at the same time or at subsequent times. This happens in cases where there are reciprocal obligations (contracts between two or more parties, treaties, conventions), or where there are many addressees (circulars, invitations, notices, memoranda), or where there are security needs (dispersal of vital records), and so on. However, we face many originals of the same document only when those originals are completely identical, as in the cases mentioned above. But, if we have a number of originals which are identical for all but the name of the addressee included in the text (think of the use of guide letters), we have as many different original documents as there are addressees. Equally, if two originals of the same document addressed to the same person

---

35. Tessier, *La Diplomatique, p. 17.*
36. Ibid., p. 18.

have a different date, they are in fact two different original documents. However, if two originals of the same document, addressed to the same person and having the same date, are sent to that person in two subsequent deliveries, the oldest document is considered to be the original, the second is qualified as *a copy in the form of original*. An example may be provided by a person asking his employer for an attestation of the kind introduced by the words "to whom it may concern." The employer sends it, but, after a while, he is asked by the same person for a second identical attestation. He copies the first and signs it, producing some-thing that is legally as perfect and enforceable as an original but lacks the quality of primitiveness that only the first attestation has.

The documents produced by computers and/or word processors might be considered a special case. Is the original the magnetic coding of the floppy disk or the printout? According to diplomatic principles, it can be either of them. If the machine-readable record, in that form, besides being the first to be produced is also complete, finished, without defect, and able to produce the consequences wanted by its author, it is the original and the printout is a copy. Otherwise, and this is particularly true for legal records which are not enforceable in machine-readable form, the printout is the original and the machine-readable record is the final draft. In practice, with computer records, as well as with all other kinds of records, one has to decide case by case which one is the original.

Further, in establishing the status of a document, that is whether it is a draft, an original, or a copy, the medium may be a consideration if it influences the enforceability of the document. Thus, in the case of photographs, the negative exists prior to the print but lacks perfection (completeness and enforceability) while the first print made from the negative is the first perfect document, that is, the original. If there are many first prints, we face the case of many originals of the same document. If many prints from the same negative or from the first print are made in subsequent times and distributed in subsequent deliveries, the first is original, the others are copies in the form of original. The same argument is valid for lithographic stones and intaglio plates which are the final draft while the numbered prints made from them are all originals. In fact, when we say "an original engraving", we refer to the print, not the stone or copperplate. Unfortunately, in common language we often use the word original to mean genuine, or first, or unique, so that we even say "an original draft" or "original sketch of a drawing." However, if all the author wants to produce is a sketch and he considers it perfect as to his intentions, it is proper to call it original. Diplomatically, this would appear to be a Contradiction but it is not, because original is used in the diplomatic sense while sketch is the technical term identifying the document artistically.

If the first perfect document is an original, what is a draft? In diplomatics, *the draft of a document is a sketch or outline of the definitive text*. It is prepared for purposes of correction and is meant to be provisional. Be it a first rough draft or a final draft ready for transcription in what will constitute the original document, it represents the creative moment in the documentation process and, because of this, has the greatest importance not only for a diplomatic understanding of that process, but also for the historical interpretation of the fact and will determining the creation of the document. However, a draft has no legal validity on its own, although such validity can be enforced by a judge in a judicial dispute, when the original either is not available or was never created and the draft is proved to be diplomatically genuine. In fact, on the basis of a certified diplomatic genuineness, a judge can declare a draft to be an authenticum, which legally means original (as defined by English law), and he can infer the existence of historical authenticity until evidence to the contrary is produced.

If the document is not an original or a draft, it is a *copy*. A copy is defined in law in rather general terms: "The transcript or double of an original writing."[37] Diplomatics makes distinctions among various types of copies. The *copy in the form of original* has already been mentioned. Then, we may have an *imitative copy* which reproduces, completely or partially, not only the content but also the forms, including the external ones (layout, script, special signs, medium and so on), of the original: a modern example is the photocopy. The probative or evidentiary value of an imitative copy is close to that of the original itself, but it does not confer on the copy legal validity in court. Normally, an imitative copy is not created to deceive, to be considered the original which it repro-

---

37. *Black's Law Dictionary*, s.v. "copy", p. 405.

duces. For this reason, it always includes elements that make the real nature of the document recognizable.

Where there is a fraudulent intention in the creation of a copy, it is a *pseudo-original*, in which the maker of the copy tries to imitate perfectly the original in order to deceive. Think of a person who copies an invitation to an event to which he was not invited in order to attend it. Legally and diplomatically, a pseudo-original is false and very often it is also historically false. In the example provided, the document on its own (without its envelope) is historically genuine (the event took place when and where indicated and invitations were sent out in that form), but it is historically false in its context because the owner was not invited. However, it would also be historically genuine in its context if the owner had been invited, had lost the original invitation, and so made a copy.

*A simple copy is* constituted by the mere transcription of the content of the original, prepared by whomever, and cannot have legal effects. This is the most common type of copy and is usually compiled as an aid to memory.

Finally, we have the *authentic copy,* which is a copy certified by officials authorized to execute such a function, so as to render it legally admissible in evidence. Also included in this category are the "inserts" (or insets), that is, the documents entirely quoted (if textual) or reported (if visual, like maps) in subsequent original documents in order to renew their effects, or because they constitute precedents of the legal act attested in the subsequent originals. A perfect form of insert is that called *vidimus*.[38] An authentic copy in general, and a *vidimus* in particular, only guarantees the con-

---

38. We have a *vidimus* when a public authority, ecclesiastic or lay, issues an "authentic act" which contains an unabridged transcription of a previous act, taking care to announce the insertion through a formula that indicates the beginning and the end of the transcription. Thus, the transcribed act is neatly individualized in the body of the new act. There are different forms of *vidimus*. Sometimes the author declares to have seen the document he transcribes, describes some of its formal characters, and affirms that it does not have any element that can diminish its legal value. At other times, a *vidimus is* a simple transcription followed by confirmation of the dispositions contained in it, the application of them to the specific case, and the addition of a new clause. Cf. Tessier, *La Diplomatique*, pp. 21-22. In English-speaking countries the formula *inspeximus is* often used in place of *vidimus*, particularly in letters patent. Cf. *Black's Law Dictionary*, s.v. "inspeximus", p.939.

formity of the copy to the original text. Thus, an authentic copy in
the diplomatic sense is also an authentic copy in the legal sense but
neither in diplomatics nor in law is it an authentic document. The
authentication provides the copy with the validity and the effects
of the original, not with its forms, and it does not influence diplomatic, legal, or historical genuineness.[39] Accordingly, if the original
was inauthentic or false in whatever sense, the copy would remain
authentic, being an *authenticated* copy of an inauthentic or false
document.[40]

Often we have many copies made either from the same original
or from copies of the same original. Now, the purpose of diplomatic
analysis of copies is to establish not only the time and context in
which each copy was made, but also the relationships among
copies of the same original. In fact, the most recent copy is not
always transcribed from the one that chronologically precedes its
Some later copies may be apographs (direct transcripts) of the
original and have thus more value for diplomatic and historical
study than previous ones which were derived from copies of the
original.

I will not illustrate the methodology involved in the identification of the sequence of copies of the same document, because, as it
stands now, it is only applicable to documents produced in the
medieval period. A new methodology for modern and contemporary material has not been developed yet. It would be useful to
investigate the feasibility of this kind of study and its relevance in
the light of the development of new legal concepts, the evolution
of documentation technology, and the changes in objectives of
scholarly research, all factors which suggest that the identification
of the genealogy of copies of the same document would be ex-

---

39. The vidimus, although part of an authentic act, does not acquire the legal
nature of that act. In fact, the public official who corroborates the new act can do so
because of his physical participation in its composition. The document transcribed
preserves the legal nature it had at its origin: if it was an authentic act, its transcription will be an authentic copy of an authentic act.

40. Among the various types of copies are registers in which documents are
reported in *extenso*. Tessier defines a register as "un livre manuscrit dans lequel une
personne physique ou morale transcrit ou fait transcrire les actes qu'elle expédie,
qu'elle reçoit ou qui lui sont communiqués au fur et à mesure de leur expédition, de
leur reception ou de leur communication." Tessier, *La Diplomatique*, p. 23.

tremely difficult and probably a sterile exercise for modern and contemporary material. But this has to be proved. What is certain is that, in the past, such an exercise contributed significantly to the establishment of the relative value of documentary sources for historical interpretation and their weight in judicial disputes. This specific function of diplomatics was part of the broader purpose I mentioned at the beginning of this section: to determine the reliability of documentary sources.

Thus, the original purpose of documentary criticism was to ascertain the historical authenticity of documents through the determination of their diplomatic authenticity (at the time referring to genuineness), with the tacit assumption that the two things automatically coincide. Such an assumption had some foundation in the seventeenth century, because the documents taken into consideration were only the solemn diplomas supposed to have been issued by royal and papal chanceries, and there was little chance that those chanceries would provide false information in that form. In time, with the broadening of the diplomatic area of enquiry to all archival documents, the coincidence of these two types of authenticity could not be assumed anymore.

Notwithstanding the noble reasons which determined the development of diplomatic criticism, and the scientific rigour of its methodology, "il n'en reste pas moins que l' enjeu de l ' expertise reste la victoire ou la défaite d'une des parties en cause," as Tessier put it.[41] Thus, until the eighteenth century, the purpose of diplomatic analysis was eminently practical and the advantages mainly, political and economic.

It is said, or at least assumed, that, when archival documents started to be considered as historical evidence and historians began to exploit them through the use of diplomatic criticism, the original purpose of diplomatics was lost. This is not entirely true. After all, Tamsin and Toustain wrote their monumental *Traité* to support one of the parties in a controversy about the documents of the abbey of Saint-Ouen de Rouen, and, until the invention of modern techniques for establishing the genuineness of a document (for exam-

---

41. Tessier, "Diplomatique," pp. 637-8.

ple, the use of chemistry), evidence was evaluated before the courts on the basis of diplomatic criticism.

It is also said that, given the evolution of the legal system, which determined the admissibility in court of types of evidence other than documentary, the establishment of diplomatic genuineness has little relevance for contemporary documents. This judgement is also too hasty. Consider the case of machine-readable records. For instance, corporations can often produce only computer print-outs as evidence in litigation. Their genuineness has to be proved and, for this purpose, foundation evidence including documentation of all the stages of a system must also be produced. Such foundation evidence has to be supported by the testimony of an expert witness vouching for the normal operation of the system or its security and authenticating the printouts created by the system. The analysis given by a witness on the operations of a computerized system is a diplomatic examination. If the witness can demonstrate that the printout has been regularly produced in a secure system, he can declare that it is diplomatically genuine and can authenticate it, that is, give it authority and legal authenticity. The historical genuineness of the printout is then inferred by the judge until there is evidence to the contrary. The same analysis could be performed by a government archivist providing inactive machine-readable records to a judge, when he would also have to document the procedures carried out during the processing and reference stages.[42] Thus, there is a continuing need for critical analysis of the genesis and forms of documents for the purposes of their admissibility as proof. Moreover, as public officials who are professionally knowledgeable of the nature of records, archivists still have an important role to play in guaranteeing authenticity of documents, and may see that role grow in significance as they acquire machine-readable records. While notaries and lawyers base their corroboration on their witnessing the formation of the document and their knowledge of the authors, the archivist bases his *a Posteriors* legal authentication on the examination of the forms *and* study of genesis

---

42. For these reflections I am indebted to Catherine Bailey, "Archival Theory and Machine Readable Records: Some Problems and Issues" (Master of Archival Studies thesis, The University of British Columbia, 1988), pp. 119-120.

of the document. In fact, in cases like that described above, diplomatic authenticity deduced from the forms of the document does not provide any reasonable expectation of diplomatic genuineness. The latter can be only ascertained through the analysis of the formation of the document.[43]

Thus, notwithstanding the technical problems presented by some contemporary documents, the different structure of their text and the specific procedures governing their creation, maintenance, and use, the basic diplomatic principles and methodology formulated for the evaluation of medieval diplomas are still valid today, and not only for the authentication function.

As I mentioned in the historical overview of the development of the discipline, in the nineteenth century diplomatics entered the category of historical sciences, because of the use made of it by romantic historicism. However, it occupied quite a minor position. As Count Simeon, French Minister of the Interior, put it in a report to the king in 1821 about the opportunity of creating the École des Chartes, "L'homme instruit dans la science de nos chartes et de nos manuscrits est, sans doute, bien inférieur à l'historien, mais il marche à ses côtés, il lui sert d'intermédiaire avec les temps anciens et il met à sa disposition les matériaux échappes à la ruine des siècles." Moreover, in 1900, in the opening lecture to the course of diplomatics at l'École des Chartes, Maurice Prou could say: "Le but des érudits français à été moins de disserter sur les règles de chancellerie et de faire de la pure diplomatique que de publier et d'utiliser les documents d'archives, en d'autre termes de donner a la pratique le pas sur la doctrine."[44]

Von Sickel is often mistakenly given responsibility for calling diplomatics an "auxiliary science of history." In fact, all he did was to introduce the teaching of diplomatics and paleography into the Austrian Institute for Historical Research founded in Vienna in 1854. Because the Institute had the function of promoting the study of the auxiliary sciences of history, for more than a century diplo-

---

43. This point is clearly made by Hugh Taylor in "My very act and deed': Some Reflections on the Role of Textual Records in the Conduct of Affairs," *The American Archivist 51* (Fall 1989), in press.
44. Quoted in Tessier, "Diplomatique," pp. 648-9.

matics lamentably came to be associated almost solely with the publication of documents of approved authenticity. Even today, most diplomatists define diplomatics as "the science which critically studies the document *in order to determine its value as an historical source,* that is, they identify the primary purposes of diplomatic criticism as historical in nature."[45]

The use of diplomatic criticism for the interpretation of historical sources is invaluable to the historian, because the examination of documentary processes and forms (which constitute the practical application of laws, regulations, and uses only partially revealed by published official sources) allows a regular verification of the discrepancies between law and actual procedure, of the continuous mediation taking place between legal-administrative apparatus and society, and of the real value of societal rules. However, if diplomatics is undeniably useful to historians of any branch of human knowledge, it is essential to archivists, who may receive from a systematic application of diplomatic methods specific benefits to their work of identification, appraisal, arrangement, and description of documents.[46]

---

45. Pratesi, *Elementi di Diplomatica Generale, p. 5.* My emphasis. The author specifies that diplomatics offers a vital contribution to history in the broadest sense, political, social, economical, administrative, linguistic, etc., through its enquiry into the administrative and legal systems in which the documents are created, and its analyses of the representation rules used. Tessier also points out that diplomatics is not a descriptive science: "Un relevé pur et simple des caractères formels et de leur variations au cours des âges ne suffirait pas. Il faut expliquer la présence des uns et l'apparition des autres en les replaçant dans leur contexte historique, juridique, social, économique, en démontrant le mécanisme de l'élaboration des actes, en scrutant l'organisation et le fonctionnement des chancelleries, le statut du personnel notarial auquel les particuliers se sont adressés ..., interroger les rédacteurs sur les moyens qu'ils ont utilisés pour exercer correctement leur métier ...", in "Diplomatique," p. 667. And Edward M. Thompson wrote: "The field covered by the study of diplomatics is so extensive and the different kinds of documents which it takes into its purview are so numerous and various...." *Encyclopaedia Britannica, I* Ith ed., s.v. "diplomatic", p.301.

46. Cencetti wrote that "diplomatics is necessary to the archivist" because it "penetrates the documentary essence and the historical formation of the papers, and determines that intimate understanding of them which is a necessary condition of their arrangement and description." Cencetti, "La Preparazione dell'Archivista," p. 285.

Diplomatics was born as a body of practical precepts, and developed as a discipline in the realm of historical studies. Once it became an historical science, it abandoned the broad area of enquiry and validity it had evinced at its origin in the seventeenth century, and transformed itself into a strictly medievalist science. However, as a consequence of the broadening of the field of archival science to include the control of active and semiactive records and the function of appraisal, archivists have rediscovered the importance of the critical study of the document and turned to diplomatics to test the validity of its principles and methods for modern and contemporary documents. The first result of this careful and laborious research is that the boundaries of diplomatics have met those of archival science, both in terms of time and place to which they are applied and in terms of methodology. Can we then talk of three diplomatics, the legal, the historical, and the archival discipline? I think not.[47] There is only one diplomatics which, when used for the purposes of another discipline, becomes one with it, just as does a metal in a metallic alloy.

---

47. Such an idea would perpetuate the concept of "auxiliary science," while it is generally agreed that all disciplines have equal scientific dignity, beyond hierarchies of importance which may be identified, that there is reciprocal trespassing among specific areas of different disciplines, and that the methods of one discipline can be used for the purposes of many others.

# Chapter 2

# The Fact, the Act, and the Function of Documents

> Record-writing must depend on some kind of interesting segregating procedure by which two things, a record and the 'world' are, first, differentiated from each other and, then, related to each other so as to make the one, ideally, 'about' the other.[1]

The revolutionary intuition of the seventeenth century diplomatists was that, if records are about the world, to gain an understanding of the world through the record requires following the same procedure which governs record-writing: first, to differentiate the record from the world; second, to relate them to each other. This intuition dramatically changed the scholar's approach to research, and gave rise to the modern philological and historical disciplines.

The only instrument which the founders of diplomatics had for understanding the world was constituted by isolated records, namely deeds of land issued by royal and imperial chanceries and preserved by various monasteries, not fonds, which were unaccessible to them because of the secrecy of archives at the time of the absolute monarchies. A small window on the world. Still, a window with a good perspective. Thus, they considered a single record, traditionally called it a document (from *doceo*, to teach), and

---

1. Stanley Raffel, *Matters of Fact* (London, Boston and Henley: Routledge and Kegan Paul, 1979), p. 17.

tried to define it according to its nature.[2] In doing so, they discovered that such nature is a whole composed of interrelated but very different groups of elements, and isolated those groups in order to analyze them. Some of the elements belonged to what the document was about, which was termed *fact*, others to the physical and intellectual makeup of the document, which was termed *form*, and still others to the procedure which brought the fact into the document, which was termed *documentation*.

At this point it was clear to the early diplomatists that to make the world speak through the document requires distinguishing "as a matter of boundaries, limits," between "the outside (what the record reports), and the inside (the record, the word)."[3] Thus, they separated the world from the document, and identified, first, what to look for (the elements of the outside), and second, what to look at (the elements of the inside). In the process of defining the elements of the outside, they recognized that the two broad groups of elements which they had termed fact and documentation, and which comprised all the world their sight could embrace, were actually two conceptually when not chronologically distinct moments: the *moment of action* and the *moment of documentation*. This recognition represents the latest and most sophisticated development of diplomatics, and the methodology of criticism derived from it is what enables us to extend diplomatic enquiry to contemporary documents.[4]

The present chapter begins the analysis of the *moment of action* by defining the fundamental concepts of fact and act, and examining the relationship they have with the document.

## The Fact, the Act and the Function of the Document in Relation to Them

Every social group ensures an ordered development of the relationships among its members by means of rules. Some of the rules

---

2. P. 41.
3. Raffel, *Matters of Fact*, p. 19.
4. Pp. 39-40.

of social life arise from the ad hoc consent of small numbers of people; others are established and enforced by an "institution," that is, by a social body firmly built on common needs, and provided with the means and power to satisfy them.[5] The latter rules are compulsory; their violation incurs a sanction or penalty. A social group founded on an organizational principle which gives its institution(s) the capacity of making compulsory rules is a juridical system. Thus, a *juridical system* is a collectivity organized on the basis of a system of rules. The system of rules is called a *legal system*. A legal system is a complex paradigm containing many divisions and subdivisions. It can be broken down into *positive law* (as set out in the various legal sources—legislation, judicial precedent, custom—and literary sources—either authoritative, consisting of statutes, law reports, and books of authority, or non-authoritative, such as medieval chronicles, periodicals, other books), and all the other *conceptions and notions of binding law* (natural law, morality, orthodox religious beliefs, mercantile custom, Roman/Canon law). Because a legal system includes all the rules that are perceived as binding at any time and/or place, no aspect of human life and affairs remains outside a legal system. For example, even the most spiritual form of love is penetrated and ruled by ethics, natural law, morality, religious beliefs, customs, and expresses itself according to them.

Each juridical system differs from all others and itself varies over time. Since the beginning of civilization, both human conduct and natural events take place within one given juridical system.[6] Within any such system, both exist as *facts*. Facts whose occurrence has not been consciously foreseen by the juridical system within which they take place are qualified as *juridically irrelevant*. Facts which are contemplated in the body of written or unwritten rules on which the system is based, that is, in its legal system, are qualified as

---

5. In a modern country such an institution is the entire organization of the state; in a primitive tribe it is the chief or the council.

6. Of course there are exceptions even to this quite general rule. In Canada, nine provinces, two territories and the federal state function on the basis of the "common law" legal system, while Quebec is organized in accordance with principles associated with the "civil law" legal system. This means that, in Quebec, the same conduct or event is subject, at different levels, to two different juridical systems, which recognize each other.

*juridically relevant.* An example may help to clarify this concept. A man receives a newly born child from the hands of the mother and holds him on his knees. Such a gesture, in the Canadian juridical system, shows affection, tenderness, and pride, but is juridically irrelevant because the system attaches no consequence to it. In contrast, in the juridical system of ancient Rome, that gesture meant that the man recognized the child as his own, and was juridically relevant because it had the consequence of legitimizing the child. This is in fact the origin of the word "genuine" (from *genua*, that is, knees), meaning true, legitimate. Therefore, a *juridical fact* is an event, whether intentionally or unintentionally produced, whose results are taken into consideration by the juridical system in which it takes place.

As mentioned before, a juridical fact can arise either from a human or from a natural cause. The first category includes such facts as a discovery or a manslaughter. For example, while heading west towards the Indies, Columbus discovered a new continent. Such a discovery, although accidental, under a principle established by the Spanish legal system, entitled the Crown of Spain, which had financed the expedition, to the property of that land. As well, if a driver accidentally kills a child running after a ball across the road, he becomes subject to penal justice. The second category includes such facts as the natural death of a person or a flood, which may be followed, respectively, by the transfer of title to the property of the deceased person, and by compensation for damage. Natural facts may produce effects contemplated by the system either of themselves or in combination with human conduct. For example, the passage of time is a natural fact that, connected with the action or omission to act of a person, has juridical effects.

Among human facts in general, the special type of fact which results from a will determined to produce it is called an action or *act*. The operation of will distinguishes an act from any other general fact. Therefore, all acts are also facts, but only facts generated by a determined will are acts. Fact is the *genus*, act is the *species*. When a juridical system takes into consideration in its body of rules not only the effects of human conduct but also the will determining it, we call that conduct a *juridical act*. In the Roman example of the baby on the knee, if the purpose of the father's conduct was to legitimize the baby, it was a juridical act; if instead it was just a gesture, although with juridical consequences, it was only a juridical fact. Moreover, the same human conduct, in the same juridical system, can be a juridical act in relation to some effects, and a juridical fact in relation to others. For example, a father's recognition of a child as his own is a juridical act inasmuch as the father intends to confer his own surname on the child, but is a juridical fact inasmuch as the father has to provide sustenance for that child (this is a juridical effect of the recognition but probably not the one primarily intended).[7] In other words, an act is a fact originated by a will to produce exactly the effect that it produces. If such an effect has a juridical nature, the will has generated a juridical act. One last example. The *fact* is: I speed in front of a school where a police car is stationed. This fact has five possible connotations:

A) Will: I want to arrive home early. Effect: ticket. Connotation: *juridical fact*

B) Will: I want to create an alibi for myself at that given time. Effect: ticket. Connotation: *juridical act*

C) Will: I want to arrive home early. Effect: early arrival (Police do not notice me). Connotation: *juridically irrelevant act*

D) Will: no will. I did not see the school sign. Effect: ticket. Connotation: *juridical fact*

E) Will: no will: I did not see the school sign. Effect: early arrival. Connotation: *juridically irrelevant fact*

Among acts in general, two categories can be distinguished: 1) acts directed to purposes immediately related to society at large, and 2) acts having effects that interest individuals or specific

---

7. The English legal system gives a definition of "act" reflecting the diplomatic concept: "An act is an effect produced in the external world by an exercise of the power of a person objectively, prompted by intention, and proximately caused by a motion of the will" (*Black's Law Dictionary*, Revised IVth ed., s.v. "act," p. 42). It is implied that the system considers as acts only those the effects of which are relevant to the system itself, that is, the juridical acts.

groups. The former category is said to take place in the sphere of public law, the latter in the sphere of private law.[8] The distinction between public and private sphere is conditioned over time by historical and political events. In fact, the social "institution" which governs the juridical system may decide at any time that some purposes which were formerly of private consequence have to be considered of public interest, or vice versa. This is because that institution, which is in a position of sovereignty, determines the criteria on the basis of which public interest is distinguished from private. For example, in Italy, in 1877, the first left-wing government decided that primary education was a matter of public concern, and gave the state full responsibility for its administration. Therefore, all acts connected to such administration, which were formerly of a private nature, became public. This type of circumstance, however, does not change the fact that, in any legal system, at any given time, there is a well understood distinction between a public and a private sphere.[9]

Within the category of private acts we further distinguish between a mere act and a transaction. A mere act is an act in which the will is limited to the accomplishment of the act, without the intention of producing any other effect than the act itself: effect and act coincide. We may consider again the above example of speeding in front of a school. If we add to our perspective the dimension offered by mere acts, that same fact has two other possible connotations:

F) Will: I want to speed because I enjoy it and I do not care about a possible ticket. Effect: I enjoy it. Connotation: *juridically irrelevant mere act*

G) Will: I want to speed because I enjoy it and I do not care about a possible ticket. Effect: I enjoy it although I get a ticket. Connotation: *juridical mere act*

---

8. Paola Carucci, *Il Documento Contemporaneo. Diplomatica e Criteri di Edizione* (Roma: La Nuova Italia Scientifica, 1987), p. 39.
9. See: Gerald Gall, *The Canadian Legal System* (Toronto, Calgary, Vancouver: Carswell Legal Publications), p. 19.

By contrast, a *transaction* is a declaration of will directed towards obtaining effects recognized and guaranteed by the juridical system. In a transaction, a person administers his/her own interests with other persons. Therefore, a transaction is an expression of autonomy of a physical or juridical person, who self-disciplines his/her own conduct in a binding way.[10]

The distinction between mere act and transaction is not operative in the sphere of public law, where acts can only assume the diplomatic configuration of transactions, because the will which generates them aims to promote the general interest of society.

Any act, to exist, must be manifested and, consequently, perceived (or at least be perceivable). This outward form of the act can be either oral or written. Diplomatics is interested in those acts which take a written form and result in documents.[11] The written form of an act can be either required or discretionary. The requirement of a written form exists in two circumstances: 1) when an act is of such a kind that can come into existence only by means of a document, and 2) when an act which takes an oral form needs a document as proof of its existence. In the former case the document is the act; in the latter, the document refers to the act. A document is also said to refer to an act when neither of the above conditions exists, and the written form is therefore discretionary.

Diplomatists have traditionally subdivided all documents into categories defined by the purpose served by their written form. In the diplomatics of medieval documents only two categories were identified. If the purpose of the written form was to put into existence an act, the effects of which were determined by the writing itself (that is, if the written form was the essence and substance of the act), the document was called *dispositive*. Examples are contracts and wills. If the purpose of the written form was rather to produce evidence of an act which came into existence and was complete before being manifested in writing, the document was

---

10. The English legal system provides a definition of a transaction which reflects the diplomatic concept: "an act ... in which more than one person is concerned, and by which the legal relations of such persons between themselves are altered." (*Black's Law Dictionary*, s.v. "Transaction," p. 1668.)

11. Once again, the term "written" is used by diplomatists in its broadest sense, as illustrated on pp. 42-43.

called *probative*. Examples are certificates and receipts. In the case of dispositive documents, the written form required for the existence of the act was defined *ad substantiam*; in the case of probative documents, the form required for providing evidence of the act was defined *ad probationem*.

The first diplomatist to approach the matter scientifically, Heinrich Brunner, resurrecting a terminology widely used in medieval documents, called the dispositive document a *charta* and the probative one a *notitia*, and made an attempt to define their respective characteristics. He considered subjective wording (with the author of the act in the first person) to be typical of the *charta*, and objective wording (with the author of the act in the third person) to be typical of the *notitia*. Moreover, studying the evolution of the most common contract in the Roman world, the *stipulatio* (agreement between debtor and creditor), Brunner generalized that documents pass in time from a probative to a dispositive value. According to this theory, while in the first periods of documentation all documents are probative, later on most of them are dispositive.[12] In very general terms, Brunner's theory is valid, at least with regard to the documents of the Middle Ages which he considered. However, it tends to blur the fundamental difference between the two categories of documents: whereas in a probative document the moment of the action precedes the moment of its documentation (for example, a birth takes place and produces effects before its entry into the birth register), in a dispositive document, the two moments are simultaneous and indistinguishable other than conceptually (for example, a sale takes place when and only when a contract of sale is completed), to the point that, in positive law, dispositive documents are usually called "acts."[13]

The inclusion of all documents in two categories can be considered valid in relation to medieval documents, notwithstanding the

12. Heinrich Brunner, *Zur Rechtgeschichte der römischen und germanischen Urkunde* (Berlin, 1880), pp. 20-21. Brunner's historical theory is endorsed by Hugh Taylor in "'My Very Act and Deed': Some Reflections on the Role of Textual Records in the Conduct of Affairs," *The American Archivist* 51 (Fall 1988), p. 459.

13. In English civil law an act is also defined as "a writing which states in legal form that a thing has been said, done or agreed." (*Black's Law Dictionary*, s.v. "act," p. 42.)

objections presented by some diplomatists.[14] In fact, nearly all medieval documents resulted from juridical acts (as defined earlier) for which the written form was required either *ad probationem* or *ad substantiam*. Furthermore, whether the will determining the act belonged to one or more persons, only one document was issued which referred to the act or put it into effect, although it could be copied or re-issued many times.

With the diffusion of education, the growing accessibility of writing instruments and materials, the development of communication systems, the increase of business activity, and the rise of complex bureaucracies, two things happened. First, people began to create documents for the purpose of communicating facts, feelings and thoughts, asking for or providing opinions, preserving memories, elaborating data, and so on. Therefore, an ever decreasing proportion of written documents came to originate from juridical acts and presented a required written form. Today, most documents are about facts, often juridically irrelevant, and their written form is discretionary. Secondly, juridical acts, and specifically those defined as transactions, began to result from a combination of related acts, juridical and non-juridical, each of which produced documents. As a corollary, many documents came to refer to the same act.

The consequences of those two circumstances for diplomatic classifications and, ultimately, for diplomatic analysis, must be examined.

The first circumstance directly concerns the diplomatic categorization of documents in relation to the function they serve. It is not possible anymore to say that written documents are either dispositive or probative, but documents of those two types continue to be created in large numbers, present the same essential characteristics identified by diplomatists of medieval documents, and are easily recognizable among all other documents. We may say that dispositive and probative documents together constitute the class com-

---

14. As an example, see Alain de Boüard, *Diplomatique générale* (Paris, 1929), p. 48.

monly, and inappropriately, called "legal records."[15] What about all the other (that is, non-legal) documents, the written form of which is discretionary? We can still use diplomatic concepts and methodology, and categorize those documents according to the function they serve, that is, on the basis of their relationship with facts and acts. If we do so, we can identify two categories which comprise all non-legal documents. The first includes the documents constituting written evidence of an activity which does not result in a juridical act, but is itself juridically relevant. We may call them *supporting* documents. The second includes the documents constituting written evidence of an activity which is juridically irrelevant. We may call them *narrative* documents. Now, what happens if we try to analyze diplomatically those two categories of documents? Inevitably, we have to adapt the methodology of diplomatic criticism to the new circumstances. In fact, in the criticism of dispositive and probative documents, we define and evaluate types of documents by their formal characteristics and formation procedures as they relate to the legal system. The legal system is a very precise reference point to which we can relate directly when examining legal documents. This is not possible when we analyze the documents of the other two categories, either because they are evidence of a continuing process which, although juridically relevant, does not result in a definite final act, or because they are evidence of a juridically irrelevant process and/or fact. However, we can still make indirect connections with the legal system, that is, we can make reference to the dispositive and probative documents issued within the same legal system in which the non-legal documents under examination have been created. We can define and evaluate types of non-legal documents by analogy, that is, by identifying first the common formal characteristics which

---

15. Such a class remains, in fact, undefined in British law. If we take the closest legal terms, that is, "legal evidence" and "legal title," we can see that they refer to anything which is admissible in court. (See *Black's Law Dictionary*, p. 1040, p. 1042.) Therefore, because anyone can attribute to the term "legal records" any preferred meaning, this writer chooses it to mean "records resulting from juridical acts for which a written form is required, either *ad substantiam* or *ad probationem*," thus closing the vicious circle: dispositive and probative documents are legal records, which are dispositive and probative documents.

they share, and second, the characteristics that each type of non-legal document has in common with a similar type of legal document.

Let us now consider in a schematic way the four categories comprising all documents:

1) documents constituting a juridical act (dispositive);

2) documents constituting written evidence of a juridical act which was complete before being documented (probative);

3) documents constituting written evidence of a juridically relevant activity which does not result in a juridical act (supporting);

4) documents constituting written evidence of a juridically irrelevant activity, whether or not such an activity will end up in a juridical act (narrative).

There is more to the above categorization than a clear-cut distinction between legal (1 and 2) and non-legal documents (3 and 4). If we reflect on the kind of documents included in each category, we may realize that the first two embrace the major part of those documents which in North America are defined as *records*, while the last two consist mainly of those documents which in North America are called *manuscripts*. Records arise from administrative activities which manifest themselves in series of acts. Those acts and their documentation are governed by written or unwritten rules of procedure, which are revealed in the forms of the records. Manuscripts, on the contrary, are the result of activities whose nature embodies a significant measure of individual freedom, which is clearly revealed in the forms of the resulting documentation. The qualification of a document as a record or as a manuscript does not depend on the nature of the creator (public or private) or on its collective or individual character (organization or person). It depends on the type of activity generating it; and because an activity is qualified by the will producing it and the effects determined by it, a document is either a record or a manuscript according to the will creating it and to the effects it is meant to produce.

Therefore, the same creator, depending on his/her purposes, may produce either a record or a manuscript.

For example, a university professor has four basic functions (teaching, research, administration, and service to the community), and a private life. In each of those five areas, he or she creates both records and manuscripts, that is, documents of all the four categories described above. As examples, let us examine the first two functions. The teaching function involves a number of activities resulting in documents: 1) giving classes, which may produce teaching notes (manuscripts, supporting); 2) giving assignments, which may produce descriptions of the assignments and evaluation documents (records, dispositive); 3) examining students, which may produce registers (records, probative) and evaluation documents (records, dispositive); 4) exchanging ideas with colleagues on course outlines, bibiliographies, etc., which may produce correspondence (manuscripts, narrative); 5) borrowing books from libraries, reserving books for students' consultation, ordering books in the bookstore, which produces a large number of certifying documents and receipts of orders (records, probative); 6) participating in the examination of theses, which creates minutes, deliberations, original copies of theses signed for acceptance, notes with the questions to be asked, and so on, that is, documents of all the four categories. The research function may involve: 1) examination of sources, resulting in research notes (manuscripts, supportive); 2) production of a draft of a book (manuscript, narrative); 3) production of the original according to contract (record, dispositive); 4) correspondence with the publisher (manuscript, supporting); 5) contract with the publisher (record, dispositive); 6) acknowledgement of the receipt of the book by the publisher (record, probative); and so on.

Having established that the typology of documents is determined by the nature of the activities generating them as qualified by will and purposes, and that, therefore, a thorough understanding of those activities is vital to an understanding of the resulting documents, we can turn our attention back to the acts.

Focusing on acts will allow us to analyse the second circumstance or condition characterizing modern documentary production: the fragmentation of juridical acts in many related but autonomous juridical and non-juridical acts, each resulting in writ-

ten documents. This condition is largely traced to the rise of bureaucracy, whose influence on documentary production has been
enormous in both the public and the private sphere. Many busi-
nesses and private organizations are in fact structured, and function, like large bureaucracies. As Stanley Raffel puts it, with the rise
of bureaucracy, the real world came "to be shaped by the very idea
of recording it .... It is not that records record things but that the
very idea of recording determines in advance how things will have
to appear."[16] Consequently, the world started to be seen as a series
of witnessable and extractable facts which, transported into the
record, became identical with the record. This evolution was determined by the circumstance that a bureaucrat, as user of the record,
wants to achieve in his/her use of the record the reality of the fact
without participating in it. Therefore, bureaucracy first divides the
world into facts, then requires the recording of them, and finally
transforms each record into a fact, into something which can be
treated as self-sufficient, ready for use. Because bureaucracy cannot
think about the record, it needs to be able to listen to the record.
"All bureaucracy can be seen as an attempt to create a method for
the reduction of contingency, imperfection, and error, an attempt
which is represented in the bureaucat-as-user's effort to reduce his
participation in the reading of the record."[17]

Bureaucracy adopts two methods for assessing the record as a
fact. The first method is indirect. Instead of deciding whether
records mirror facts, bureaucracy decides whether record-writers
are reliable. If the writer is reliable, the user can identify him/her-
self with the writer, that is, with the witness to the fact. To be able
to rely on record-writers requires controlling them in a number of
different ways. Raffel identifies four ways: 1) restricting the privilege of record-writing to professionals (of course, record-writers

---

16. Raffel, *Matters of Fact*, p. 48. Writing about the documents produced by
bureaucracy, this writer will use the term "document" and "record" interchangeably.
In fact, all bureaucratic activities are aimed to the production of juridical acts.
Therefore, all documents bureaucratically created are records, even if the opposite
is not true. The term "bureaucracy" is used here to refer to any kind of administrative
apparatus presenting a structure similar to that of government bureaucracy, be it a
church, a business, a university or other.

17. Raffel, *Matters of Fact*, p. 78.

are those who sign records and/or are responsible for them); 2) imposing sanctions on record-writers by requiring signatures, so that the bureaucrat has a record anyway, either a record of the fact or a record of who failed to report the fact; 3) instituting procedures, that is, giving responsibility to each writer for reporting only a portion of a fact, and/or increasing the number of those who report the same fact, so that what their records will have in common will be the true fact; 4) making different purposes concurrent, that is, making the same record serve a number of different users: instead of the number of writers, the size of the audience is increased, so that the writer cannot tailor the message to the audience.

The second method for assessing the record is direct. Rather than being concerned with the truthfulness of the record, bureaucracy focuses on its completeness. Records can be assessed in terms of standards other than their effectiveness in mirroring facts, that is, they can be assessed in terms of forms. This evaluation amounts to redefining the *record as a visible fact at which the user is present*. If a record possesses all the various bureaucratically necessary forms and those forms are complete, the user can achieve complete passivity and treat the record as a thing which is showing him/her what it is. Completeness is the major standard in terms of which records are actually assessed. Any manual, directive, or circular related to record-making emphasizes, not that records should be truthful, but that they should be complete. Completeness is the bureaucrat's way to the real. How? Let us take as examples two elements common to various record forms: signature and date. By requiring a signature, bureaucracy asks writers to declare by signing that their records mirror the facts. The declaration that a record is adequate becomes the fact for the user. The signature gives responsibility to the writer for his/her words; therefore the user does not need to check the record against the fact, because the signature shows and legally establishes where the responsibility lies. The signature is the fact. By requiring the indication of the place and time in which a record is written, bureaucracy transforms the record into the fact, because the mention of a topical and/or chronological date captures the relation between writer and fact, and this relationship becomes one of the things the record speaks about: a fact belonging to the past can be known by the record-user if the relationship between the person who writes about it and the fact itself is localized in space and time.[18]

The facts bureaucracy deals with are of any kind, but the facts bureaucracy is directly involved in are of a very special kind; they are juridical acts directed to the obtainment of effects recognized and guaranteed by the system, that is, they are transactions. The necessity of examining the characteristics of bureaucratic transactions for an understanding of records bureaucratically produced is made clear by Raffel's analysis of the relationship between bureaucracy and records. It can be further strengthened if we consider a few words written in a different context, for different purposes: "Records are recorded transactions. Recorded transactions are information, communicated to other people in the course of business, via a store of information available to them."[19] This statement defines records only indirectly as information. Their being information descends from the fact that they are recorded transactions. Moreover, their being recorded transactions qualifies the type of information which they are. They are not just recorded information, but conveyed information. The author of the definition, the United Nations Advisory Committee for the Coordination of Information Systems, is so conscious of its implications that it feels the need to explain further: "This definition ... is consistent with the concept that a record is created by an official action of receiving or sending information. Both paper based records management and electronic records management must distinguish between the hour-to-hour or day-to-day changes in a draft of an official document and records

---

18. Ibid., pp. 48-116.

19. United Nations Advisory Committee for the Co-ordination of Information Systems, Technical Panel on Electronic Records Management (TP/REM), Electronic records guidelines: a manual for policy development and implementation (Fifth Session of ACCIS, 18-21 September 1989), p. 10. These guidelines constitute the first part of the final report of the ACCIS Technical Panel on Electronic Records Management (TP/REM). The full report includes three other sections, namely: "Glossary for electronic archives and records management"; "Standards paper (Integrated systems management of official records and documents in United Nations organizations: a requirement of the 1990s)"; "Report of the TP/REM survey on E* management (Paper on technical integration)." The guidelines, as well as the other three sections of the full report, can be requested at ACCIS Focal Points (Geneva and New York).

sent or received by the organization. In both situations making an entry in a bookkeeping journal, a case file, a database, or even a 'memo to the file' is creating a record even though the information is not 'sent,' because others are intended to receive this communication at a later date. Each system must distinguish official from purely private information; thus jotting a note about an expenditure or change of address on a loose slip of paper or in an electronic memo pad to remind ourselves to make such an entry at a later time is not a record-transaction, and hence, not a record."[20] This statement has enormous implications. The United Nations is a bureaucratic organization, which like any organization functions by means of transactions. These transactions take place by means of documents which, as Raffel pointed out, must be reliable and complete, so that they can be identified with the transactions they are about. Therefore, documents which are reliable and complete, that is, able to convey information, capable of being used in a transaction, and of reaching the purposes for which they have been produced, *are* transactions. This species of documents are called records. Records are recorded transactions. As a consequence, recorded information, or documents, which do not present the above characteristics, are not records. They are still documents, though, and may be very significant documents, because they reveal the creative process of producing records, that is, of carrying out a transaction. This writer does not believe that the committee, making the distinction between them and the records, aims to re-create the medieval dichotomy between the archives-treasure, made of chosen documents, mainly dispositive and probative in original form (think of Le Trésor des Chartes), and the archives-sediment, made of interlocutory documents, mainly drafts and notes, produced constantly and progressively in the conduct of affairs. Referring to a very different operation, the committee is saying to a bureaucratic organization that, to acquire and maintain control of its documentary material, it is vital that both its record creators and its record managers make a distinction between the documents resulting from a *procedure* and those resulting from a *process*. If a procedure is the body of written or unwritten rules whereby a

---

20. Ibid.

transaction is effectuated, and comprises the formal steps to be undertaken in carrying out a transaction, the documents resulting from it are one with the transaction, and must be identified as such since their creation, so that they will not be confused in the future with those generated by processes. In fact, a process is a series of motions, or activities in general, carried out to set oneself to work and go on towards each formal step of a procedure. The documents resulting from a process are preparatory, incomplete, they are the instruments necessary to set the stage. They are not meant to be communicated, and may be as precious to the scholar as they are irrelevant to the bureaucracy. The committee calls records the products of procedures and, in the manual, does not consider the products of processes. Undoubtedly, there are good reasons: the work is directed to an organization which, because of its bureaucratic nature, cannot be interested in non-records; the guidelines refer to the management of electronic documents, among which the non-records are often just scattered pieces of unrelated information; and the focus is on current documents, both for the needs of the organization and for the practical problems presented by the specific medium.

For diplomatic purposes, we will continue to call non-records "written documents," as defined by diplomatics. We will not call them just "recorded information" as opposed to "communicated information," because diplomatics does not focus on content, but on context and forms. Therefore, "documents" are the *genus*, "records" are a *species*.

In order to continue the verification of the applicability of diplomatic principles and methodology to modern and contemporary documents, and, at this particular point, to documents created by bureaucratic organizations, it is necessary to review the characteristics of bureaucratic transactions. For reasons of simplicity, the transactions will be called "acts," and their documentary result will be called "documents."

Bureaucratic acts can be classified into various groups:

1) when the power of accomplishing the act is concentrated in one individual or organ we have a *simple act*; if the organ comprises a number of individuals we have a *collegial act*. In

the latter case, just as in the former, the will to produce the act is one will, because the collective will of the single members is manifest in one deliberation with one purpose;

2) when the power of accomplishing the act belongs to two or more interacting parties (individuals, public bodies, states, state-and-individuals), we have a *contract*. Notwithstanding the difference in motivation and interests between the parties, their wills converge in one, aimed at producing the one act;

3) when the accomplishment of the act depends on many manifestations of will either of the same individual or organ (at different times), or of many different individuals or organs, we may have:

   a) *collective acts*, produced by the identical wills of different individuals or organs, and resulting in one document (for example, a circular signed by a number of ministries);

   b) *multiple acts*, produced by the will of the same individual or organ but directed to different individuals or organs, and resulting in one document (for example, a document giving merit increases to a number of employees);

   c) *compound acts*, composed of many different acts produced by the same individual or organ or by a number of individuals or organs, but all essential to the formation of some final act of which they are partial elements. The partial acts may concern the same or different subjects and may respond to convergent or contrasting interests, but each results in documents, which are all necessary to the formation of the final document: the final product of the compound act.

Compound acts can be further divided into:

   - *continuative acts*, when the same individual or organ needs to manifest the same will more than once in order to produce the final and definitive act, so that the partial

acts constituting the compound act are all identical, but
the documents resulting from them are not (for example, a City Council's three subsequent deliberations of
the same by-law);

- *complex acts*, when different individuals or organs, which
may have different motivation and interests but pursue
the same function, produce a number of simple acts
having the same content, all necessary to the accomplishment of the final act (for example, the series of
approvals necessary to the appointment of a Dean);

- *acts on procedure*, when the final act derives from a series
of different acts (which may be simple or compound,
collegial or collective, in sequence or parallel), produced
by a number of different individuals and/or organs,
which have equal or different motivation or interests
and accomplish different functions. However, all these
partial acts have the common aim of making possible
the accomplishment of the final act (for example, the
procedure needed to create a new curriculum of studies).[21]

The above classification of acts shows a panorama of the "world"
very different from the one seen by the diplomatists looking
through the window of the medieval record. If we imagine this new
world (a modern juridical system) as a city, we can still see in it, as
in medieval times, quite a number of one-family homes (simple
acts), homes in which the families of the sons share multi-level

---

21. Carucci, *Il Documento Contemporaneo*, pp. 42-43. The formation process of
documents resulting from compound acts will be examined in some detail in the
fourth chapter of this book.

The creation, in this classification scheme, of a distinct category for "acts on
procedure" may appear to be contrary to the previous suggestion that all bureaucratic transactions are based on procedure. However, in the context in which that
suggestion was made, the term "procedure" is used in its general meaning, while
here it is used in its technical meaning of machinery set up by legislative means
(legislation, regulations, directives) for carrying on a given transaction.

space with that of the father (collegial act), duplexes or villages of townhouses built by a consortium of families (contract), apartment buildings (collective acts), homes for poor or old people (multiple acts); but we also see bus terminals, railway tracks, airport radars, that is, a great number of constructions which cannot be understood on their own (partial acts) because they are only one element of a much larger whole, a transportation system (compound act). Now, it is true that a one-family home cannot tell us everything about the city (the juridical system) of which it is part, but it can reveal a lot about the family living in it (the will, the effects), its environment (the "institution"), and its social rules (the legal system). What can railway tracks reveal? Well, among other more specific things, they reveal that they are part of a greater whole, and this fact compels us to look for the rest.[22]

The writer chose to compare a juridical system with a city because concepts may be more understandable if made visible. However, such a simile is not entirely appropriate, because acts need not result in visible, concrete things. Herein lies the only significant difference between the juridical system described by diplomatists of medieval documents, and the modern and contemporary one. The categories of acts which bureaucracies since the Renaissance accomplish, in both the public and the private spheres, have always existed. However, in the Roman Republic, the Carolingian Empire, and medieval city-states, each of those acts, whether simple, collegial, contract, collective, multiple or compound, produced one and only one document. It often preserved within its text the memory of partial acts contributing to the formation of the final act, and of their determining wills. This could happen because the partial acts were oral: a written form was not required for them to exist. The authority or reliability, in Raffel's

---

22. Of course, a large part of the greater whole will always escape diplomatic examination, no matter how thoroughly it is conducted. Broad political, ethical, intellectual, and generally societal questions can find only partial answers if we limit ourselves to the study of the formation and inner constitution of documents. This was strongly felt when diplomatics was first applied, and is the reason why, from diplomatics, all historical and philological sciences developed, including archival science. This is also the reason for which diplomatics constituted the foundation of, and later was fully incorporated in, individual historical and philological disciplines.

terms, of the author of the document manifesting the final act was such that a simple mention in that document of related partial acts in oral form was a sufficient proof of their existence.

Therefore, the diplomatic concepts referring to juridical and legal systems and to facts and acts, the diplomatic theory of a distinction between the moment of action and the moment of documentation, and the diplomatic principle that each document is linked by a unique bond to the activity (be it fact or act, juridically relevant or irrelevant) producing it, a bond qualified by the function served by the document, are still valid and able to guide the diplomatic analysis of modern and contemporary documents.

From the continuing validity of the conceptual foundations of diplomatics descends the continuing validity of its methodology for the analysis of documents bureaucratically produced. However, when applying the methodology, we must be prepared to deal with consequences unforeseen by the founders of the discipline, that is, we will have to find ways of taking that methodology much further.[23] The early diplomatists identified the elements of the outside world we have to look for in a document, with the assumption, based on the then known reality, that it is possible to go directly from the document to the entire fact generating it. Their methodology presupposed that there is a bilateral relationship between each document and the fact it is about, so that if a fact, (A), is manifested in written form, the document resulting from it, (B), will guide us directly to the fact: A—B—A. This direct, exclusive, bilateral relationship exists only for a limited number of documents in modern bureaucracy. For example, we can see it in a last will or a receipt, that is, in the purely private parts of bureaucratic transactions. If a man expresses in writing his wishes about the destination of his properties after his death, the resulting document mirrors all the fact: last will (which is a juridical act). But, what about the fact: inheritance? As well, if one pays a membership fee to the Association of Canadian Archivsts, and obtains a receipt, that document mirrors all the fact: payment (again, a juridical act). But, what about

---

23. Remember that, in relation to the documents of categories n. 3 and n. 4, that is, to the *species* of documents commonly called "manuscripts," the writer talked about adaptation of methodology. Instead, in relation to bureaucratically produced documents, she does not think that diplomatic methodology should be adapted.

the fact: becoming an ACA member? Therefore, applying diplomatic methodology to modern and contemporary documents, we will find ourselves faced with multilateral relationships, in which each single fact manifests itself in a fragmented documentary form, and each document guides us not only to a small portion of the fact it is about, but, possibly, to a chain of other documents and/or facts. The bond which links a document to the activity producing it is still unique, but is probably not the only relationship that such a document has. Remember those constructions in the modern city, the meaning of which could be captured only by examining their complementary parts? The single document of the Middle Ages might be the dossier of modern times. We will examine this possibility at the end of this book on diplomatics.

Readers of this chapter should consider the fact that, in the near future all the professional work of archivists may focus on the subjects which have been discussed: the juridical system, the fact, the act, the will, the effects. Can you hear the echo of Hugh Taylor's words? "Their impact [the impact of letters] is not assessed ... on the basis of content but on the action of writing ... Increasingly, the act or decision which informs the conduct of affairs grows closer in time to the document that records it ... Electronic communication ... can become a continuous discourse without trace, as both act and record occur simultaneously with little or no media delay or survival. Words once again become action oriented."[24] However, because there can be no acts without actors, the third chapter of this book will discuss the persons concurring in the formation of acts and documents, and the nature of documents in relation to those persons.

---

24. Taylor, "'My Very Act and Deed'," p. 468.

# Chapter 3

# The Persons and the Public and Private Nature of Documents

"This is very true: for my words are my own, and my actions are my ministers."[1]

The statement of Charles II made in response to his premature epitaph written by Lord Rochester encapsulates one of the most significant problems encountered by historical enquiry on sources: the discrepancy which may exist between historical facts and juridical facts.

The distinction between historical truth and juridical truth is one of the pillars of diplomatic criticism. This distinction does not imply that the two entities are necessarily in conflict or that either of them constitutes the higher truth; rather, it means that they belong to different logical categories, and that a direct connection between them would lead to arbitrary and perhaps unwarranted conclusions. When juridical facts manifest themselves in documentary forms, they constitute the documentary truth, which can be revealed by an analysis of the formal elements of documents. On the contrary, when historical facts enter documents they manifest themselves in their informational content, in the message expressly

---

1. These words are the reply of Charles II (1630-1685) to Lord Rochester's premature epitaph: "Here lies our sovereign lord the King/whose promise none relies on;/He never said a foolish thing,/Nor ever did a wise one."

conveyed by the documents, and an examination and interpretation of that message is necessary to ascertain the historical truth. We can say that a document *is* a juridical fact and it *is about* historical facts. In the case of Charles II, the documentary truth is that he was the author of all his actions, which were done in his name and by his command, and which could achieve their effects only because they were actions of the King. The historical truth that Charles II did not know about many of those actions, or explicitly opposed them, has been revealed by historians' interpretations of information conveyed by various sources, but has no bearing on the effects of the actions themselves within the juridical system of the time.

The distinction between historical truth and documentary truth corresponds to the distinction postulated by diplomatics between the *moment of action* and the *moment of documentation*.[2] Any action or *act* originates from at least one will: are the will producing an act and the will producing the related document one and the same? If not, how do they differ? How are they related? Which one has effects in a given juridical system? As there is no act without an actor and no document without an author, this chapter will answer the above questions indirectly, by introducing the persons who intervene in the creation of a document, and discussing the nature of documents in relation to them.[3]

## The Persons Concurring in the Formation of a Document

Persons are the central element of any document. We identify, acquire, select, describe, communicate, and consult documents largely in relation to the persons they come from, are written by, directed to, concerned with, or have effect on. Persons have been the focus of diplomatic theory since its first formulation, and their identification has constantly presented problems to diplomatists, because persons are the element most tightly linked to juridical

2. P. 60.
3. For the purposes of this study, the term "person" is used in the sense of "juridical person," as defined in note 20 on p. 42.

conceptions and systems, and the most influenced by their subtlest variations.

In a diplomatic context, as well as in the legal one, persons are the subject of rights and duties; they are entities recognized by the juridical system as capable of or having the potential for acting legally. Just as collections or successions of human beings (juridical persons) derive their qualification of persons from a legislative authority, individual human beings are not automatically persons because they are human, but become such if rights and duties are ascribed to them. Every full citizen is a person; subjects may be persons; but human beings having no rights are not persons. The person is the legal subject or substance of which the rights and duties are attributes. Because a person is a character in virtue of which certain rights belong to a human being or to a collection or succession of human beings, and certain duties are imposed upon him or it, one human being or a group of human beings may unite many characters as, for example, the characters of father, husband, president of a club, or the characters of professional association, publishing company, and investor. The term person derives from the Latin word *persona*, which means mask, character, part. With respect to their own juridical system, human beings are what they represent at any given time; with respect to the documents they create and receive, human beings are what they impersonate in each given document; they coincide with the part they play. It is not a coincidence that in legal documents the concerned persons are often called parties.[4]

Diplomatic theory posits that, for a document to come into existence, the concurrence of at least three persons is necessary: the two protagonists of the action and the writer of the document. In some documents, two of those persons, or all three, are one and the

---

4. The Italian writer Luigi Pirandello argues that in our daily life too, we wear as many masks as the relationships we engage in (see *One, Noone, Onehundred*), and that we have no existence outside our network of formal relationships (see *The late Mattia Pascal*). Crying on the grave of his mother, Pirandello mourned the death of one of his own characters: he could not be a son anymore! His conclusion seems to be that, because there are as many truths as the parts we play and each of them is nothing else than a mask, the one and real truth about ourselves transcends us and exists only in front of God.

same entity or juridical person, individual, or group, but such entity plays different roles, and diplomatics focuses on the roles, the parts, the masks.

The terms used by diplomatics to designate the three necessary persons are: *author*, *addressee*, and *writer*. There may be other persons intervening in the creation of a document, such as witnesses, registration or authentication officers, or clerks. These persons do not participate in the creation of every document, and diplomatics designates them by the function they accomplish in the document under examination.

The *author of a document is the person(s) competent for the creation of the document, which is issued by him or by his command, or in his name* (e.g., the testator in a will, the King in a letters patent, the university in an honourary degree diploma). The name of the author may appear in impersonal form in the heading of the document (e.g., The University of British Columbia), or at the beginning of the writing (e.g., George The Third, by the grace of God ...), or in any section of it, or in the subscription of the document.

Usually, the author of a document coincides with the author of the act put into being or referred to by the document, because the person whose will has given origin to the act documented tends to be also the person competent for the creation of the related documentation. Thus, a jury is author of a verdict, a university of a letter of appointment of a professor, a Queen of a patent for an invention, a city of its by-law, a student of his examination paper, and so on.

Sometimes, the person competent to document an act is different from the author of the act itself. This is more common in the sphere of private law, when acts accomplished by private persons are documented by public officers, lawyers, or notaries. For example, in a contract of sale created by a lawyer, the author of the act is the seller, while the author of the document is the lawyer. Both French and Italian diplomatists have recognized the conceptual distinction between author of the act and author of the document, but only German diplomatists have coined distinct terms for them: *Urheber* (author of the act), and *Austeller* (author of the document).[5] The

---

5. See Alain de Boüard, *Manuel de Diplomatique Française et Pontificale. Diplomatique Generale* (Paris: Editions Auguste Piscard, 1929), p. 40; and Alessandro Pratesi, *Elementi di diplomatica generale* (Bari: Adriatica Editrice, n.d.), p. 26.

wording of the document and a knowledge of the juridical system within which the document is created are the keys to identifying the author of the document. Chanceries or offices entrusted with the documentation function can never be authors of documents in so far as they act in the name of the person they serve.

It has been suggested by some archivists that, with electronic records, authorship should be partially attributed to systems designers, who provide the systems with "knowledge" necessary to the creation of records. This writer strongly disagrees with this proposal because, as Margaret Hedstrom puts it, electronic systems do not possess more knowledge than a code of administrative procedure used in a complex records office: they both are "embodiments of human choices," adopted because they satisfy the needs and requirements of those whose *will* determines the creation of records.[6] For example, there is no difference between entering the amount of a sale into a spreadsheet with the understanding that the system will calculate the related taxes and totals according to the instructions received by its designer, and providing a records office with the same initial information with the understanding that an accountant will make the calculations and a secretary will compile the final document according to the established administrative procedure. Briefly, if I write a letter using WordPerfect, the author of the letter is not WordPerfect. This does not mean that diplomatics does not consider it important to know about the electronic systems used for records creation, and how they work. It means that diplomatics does not consider them to be "persons," but instruments for the realization of the will of persons, and, as such, part of the broad area of procedures and processes for records creation, which will be treated in the fourth chapter of this book.

The *addressee of a document is the person(s) to whom the document is directed*. The name of the addressee may appear on the top of the document, or in its text, or at the bottom, or on the *verso*. The addressee of a document usually coincides with the addressee of the act put into being or referred to by the document (e.g., the heir

---

6. Margaret Hedstrom, "A Research Agenda for Automated Records," speech given at the School of Library, Archival and Information Studies, The University of British Columbia, 6 March 1990.

in a will, the beneficiary in a concession, the appointee in a letter of appointment). There is no document without an addressee because documents result from actions and any action falls on somebody. An action may be directed to an entire collectivity, and in such a case the addressee of the related document may be all the people, or a social, ethnic, or religious group, and so on. An action may be directed to its author, who will be at the same time author and addressee of the related document (as in a personal diary, or in a cheque written to oneself). An action may be bi- or multilateral, that is, may involve reciprocal obligation of two or more parties: in such case each party will be author and addressee of the related document.[7] Finally, if an action is unilateral, that is, if it originates from the will of one person (either an individual or a body), the addressee of the document may be either active or passive, that is, may need to be notified of the action or not. For example, the offer to buy a house not only needs to be documented in writing, but also the resulting document needs to be communicated to the addressee for the action to take place; on the contrary, a last will, for which a written form is equally required, does not need to be communicated to the addressee for the action to be concluded (the action in question here is the last will-action, not an inheritance-action).

Sometimes, the addressee of the document is different from the addressee of the action. An example is offered by documents like licences, permits, and patents, the most solemn of which contain the expression "To all to whom these presents shall come," where the addressee of the document is whoever is concerned with the fact attested in it, while the addressee of the act is the person whose name appears in the document. The same is true for all forms of registration: in a land property registration the addressee of the action is the owner of the land, while the addressee of the document is the city holding jurisdiction over the land.

The *writer of a document is the person(s) responsible for the tenor and articulation of the writing*. The name of the writer, often accompanied

---

7. Notwithstanding this fact, for purposes of description, diplomatics has established a convention by which the first party mentioned in the document appears as the author, and the second and other parties as addressees. Thus, in an agreement for a sale, even if conceptually seller and buyer are at the same time both author and addressee, the seller is considered to be the author and the buyer the addressee.

by his qualification, usually appears at the end of the document, and assumes the form of a subscription, but sometimes we encounter it on the left margin or on the top left of the document.[8] The writer of a document may coincide with the author or one of the authors of the document (if I write to somebody in my private capacity I am the author and the writer of the letter); or be a delegate of the author (I may ask a colleague to answer my correspondence while I am absent from the office, giving general directions on the content of the answers: I would be the author and my colleague would be the writer of those letters); or his representative (Alured Clarke, Lieutenant-Governor and Commander-in-Chief of the Province of Lower Canada, was the writer of all the letters patent authored by George the Third and related to that Province); or one of its officers, when the author is an abstract entity, such as a state, a city, a university (a mayor is the writer of the documents authored by the city he or she serves); or one or more of his members, when the author is a collective entity, such as a board, a committee, a task force (the chair of the Education Committee of the Association of Canadian Archivists would probably be the writer of the letters that the Committee as the author sends to the ACA Executive as the addressee). The writer is not a clerk or a secretary, because these individuals are not "persons" with regard to the documents they compile: they are not competent for the articulation of the discourse within the documents, unless they have delegated authority for it (a clerk may be competent for writing the entries in a register, but usually the responsibility belongs to an officer, who may have the title of Secretary, but with a meaning quite different from that which was referred to).

It may happen that the subscription of the writer is followed by another subscription, which may be qualified by the word "countersigner" or "secretary."[9] This signature has the special function

---

8. The subscription or signature may be an autograph or not, the difference between the two being that the autograph needs to be handwritten by the very hand of the signer, while the subscription or signature does not. For example, a telegram is subscribed or signed, not autographed.

9. In diplomatics, *qualification of signature* is the term used to refer to the title of the signer, or his position, or his function with respect to the subject of the document. For instance, an archivist may have his signature followed by the words "state

of validating the physical and intellectual form of the document and of guaranteeing that the document was created according to the established procedure and signed by the appropriate person. The countersigning person assumes responsibility only for the regularity of formation of the document and for its forms; that is, not for its content, and not for the wording chosen to express the content, but for the presence in the document of all the elements required for its effectiveness, such as title, date, address, conditions of the transaction, signature, stamp, mention of taxes paid, and the like. Sometimes, the signature of the countersigner is not expressly qualified, or is qualified with the official title of the individual in the administrative hierarchy (for example, the signature of the City Clerk following that of the Mayor in a by-law is often not qualified by the word "countersigner," but by the title "City Clerk"): in such case it is essential to identify conceptually the function of the signature, and not to confuse the countersigner with the writer or with the author. Some documents, particularly solemn dispositive documents, present two countersignatures, one just under, or at the side, of the signature of the author and/or writer, and one much below, often at the left corner of the page. This second countersignature has the function described above, of procedure and form control, but the first contersignature has a very different function and sometimes is not accompanied by the second. It appears more frequently in documents whose author does not coincide with the author of the act put into being or referred to by the document itself, and belongs to the author of the act (e.g., the parties in a contract authored by a lawyer), or to the author's representative or, if the author of the act is an abstract entity, to one of its officers. It constitutes a declaration that the content of the document corresponds perfectly to the will of the authority, and it is easily distinguishable from the other type of countersignature because the signing person is qualified by a title which shows direct responsibility for the type of act mentioned in the document (e.g. commis-

---

archivist" or "chairman ICA Education Committee," or "member, parents association." Whatever the words used, they qualify the preceding signature and give a place to the subscriber in the context of the document. This intrinsic element of intellectual form will be analyzed in the fifth chapter of this book.

sioner of patent, legal counsel, managing director), and not for the functions associated with records or documentation offices.

As pointed out at the beginning of this exposition, many other persons may concur in the formation of a document, such as witnesses, records officers, and records clerks. Such persons have many different functions with respect to the documents with which they are involved. For example, the signature of a witness may serve to confer solemnity on a document, or to authenticate the signature of the author (either of the act, or of the document, or of both), or to validate the content of the document, or its compilation, or to affirm that an act for which both oral and written form are required, such as an oath, took place in his presence, and so on and so forth. However, because those persons are usually clearly identified as to their function in the various documents in which they are named, and cannot be confused with the three persons essential to the formation of a document, their functions will, in the fifth chapter of this book, be discussed in the context of the extrinsic and intrinsic elements of documentary forms, and of the concepts of validation and attestation, authentication, and registration. Furthermore, differently from the three essential persons, witnesses and records officers or clerks do not have direct influence on the nature of the document in which they appear, but only on its effectiveness; therefore they are not relevant to the purposes of this chapter.

When many persons intervene in the creation of a document and/or subscribe it, it may be difficult to identify their individual roles, but it is important to do so, because those roles have a bearing not only on the nature and effects of the document but also on the way the archivist will deal with it (e.g., in arrangement and description). In the identification of persons, there are two key words the diplomatist needs to keep in mind. The first is *responsibility*. Responsibility is the obligation to answer for an act. The second is *competence*. Competence is the authority and capacity of accomplishing an act. The two entities may or may not converge in one. In our juridical system, when responsibility is given to a juridical person for a specific function entailing a number of defined acts, that same person is often given also the competence for that function, and vice versa. In fact, in the modern administrative context, competence may be defined as the area of responsibility within a function.[10] However, it is not so in all juridical systems. While in a constitutional monarchy, for example, the sovereign is competent and responsible for some functions attributed to him by the constitution, in an absolute monarchy, the sovereign is competent but not responsible for all his actions. Therefore, the two concepts must be used separately in documentary criticism.

When looking for the persons concurring in the formation of a document, we have first of all to look for competences and then for responsibilities, and ask, in relation to the juridical system of the time and place concerned:

1. Who was competent to accomplish the act put into being or referred to in the document; that is, who had the authority and the capacity to accomplish it?
2. In whose name did the competent person act?
   a) If in its own, which part was played (e.g., engineer, friend, husband, investor, coach of a team?)
   b) If in the name of another person, did the responsibility for the act fall on the competent person or on the person represented?
3. Who was competent to issue the document?
4. In whose name did the person issue the document?
   a) If in its own, which part was played?
   b) If in the name of another person, who was responsible for issuing the document?
5. Who was competent to articulate the writing?

---

10. Function and competence are a different order of the same thing. Function is the whole of the activities aimed to one purpose, considered abstractly. Competence is the authority and capacity of carrying out a determined sphere of activities within one function, attributed to a given office or an individual. For example, the archival function consists of all the activities whose aim is the preservation and communication of archival material. Within this general function are sub-functions like appraisal and arrangement. Each archival institution has the competence for fulfilling a defined portion of the general function (e.g., preservation and communication of the archival material created by a determined provincial government), and each archivist has the competence for accomplishing a defined portion of the subfunctions (e.g., the reference archivist of a provincial archives makes available the material preserved by that institution; or, the archivist competent for the judicial records of a given province, acquires, selects, arranges, and describes those records). Therefore, a competence practically coincides with a portfolio. While a function is always abstract, a competence must be attached to a juridical person.

6. Who was competent to establish the formation and forms of the document?
7. To whom was the act directed?
8. To whom was the document directed?

In many cases, the answers to these questions are very simple and we need not even ask them. However, when we encounter documents created in a bureaucratic context, particularly if we are not familiar with that context, much research is necessary to answer these questions and identify the persons necessary to the existence of a given document. This may involve a study of intellectual and legal history, of juridical and administrative conceptions, of structures, organizational and procedural rules, and, of course, an analysis of the other documents created in the same context. The earlier diplomatists did not have available to them all the instruments and knowledge available to us today. Therefore, their sources consisted mainly of the documents created in the same context in which the document under examination was produced. This is the reason they brought together in published volumes documents issued by the same juridical person (Emperor, King, Pope, etc.) and preserved by the various addressees.

At this point we may try to identify the persons involved in the creation of two documents produced within two different juridical systems by answering the questions listed before.

Let us consider first the letters patent of *dedimus potestatem*, which apppears in Figure 1.

| Questions | Answers |
| --- | --- |
| 1 | George the Third |
| 2 | George the Third |
| 2a | King of Great Britain, etc. |
| 2b | NA |
| 3 | Alured Clarke, Lieutenant-Governor of the Province of Lower Canada |
| 4 | George the Third, King of Great Britain, etc. |
| 4a | NA |
| 4b | George the Third |
| 5 | Alured Clarke |

6                    George Pownall, Secretary and Registrar of the
                     Province of Lower Canada
7                    George Pownall and Jenkin Williams, Esquires
8                    George Pownall and Jenkin Williams

Persons

Author of the act: George the Third
Author of the document: George the Third
Addressee: George Pownall and Jenkin Williams, Esquires
Writer: Alured Clarke (signature on the top left, where the King
himself would sign)
Witness: Alured Clarke (initials at the bottom)
Countersigner for procedure and forms control: George Pownall

The second document belongs to our juridical system, and is
regularly created by a type of administration so familiar to us that
we do not even need to visualize it as one specific document, but
we may think of it as a typical documentary form: a letter of
appointment of a professor in a Canadian university, shown in
Figure 2.

Questions        Answers

1                Various levels of the University hierarchy, as
                 established by the regulations of the University
2                The University
2a               NA
2b               The University
3                The President of the University
4                The University
4a               NA
4b               The University
5                The President of the University
6                The Office of the President
7                The appointee
8                The appointee

Figure 1: Letters Patent of George III, signed by Alured Clarke. *Courtesy: National Archives of Canada, Civil and Provincial Secretary Lower Canada ('S' Series) 1760-1840, RG 4, A1, volume 54, page 17910. Copy negative number: C-135892.*

THE UNIVERSITY OF BRITISH COLUMBIA

OFFICE OF THE PRESIDENT
107 - 6328 Memorial Road
Vancouver, B.C. V6T 2B3

March 13, 1987

Name:
Faculty:
Department:

The Board of Governors has authorized your
appointment as indicated.

Title:        Assistant Professor

Term:

Pay:

Enquire at Faculty and Staff Services, Room 300, General
Services Administration Building, 228-4541, for
information on benefits (Retirement Plans, Group Life, and
Disability Insurance, Medical and Dental care).

This appointment is subject to the provisions of one of the agreements
on Conditions of Appointment - (Faculty; Program Directors of Centre
for Continuing Education; Professional librarians).

This appointment is subject to the conditions on the reverse and the
other provisions set forth in the Faculty Handbook and in the Manual
of Policies and Procedures of the University, insofar as they may be
applicable. Pages C-1 and C-2 of the Faculty Handbook indicate which
classes of appointment carry an implication that the appointee will be
considered for reappointment at the end of the specified term.

Travel and removal allowance in accordance with PeB-12.

Appointee

PRESIDENT

Figure 2: Letter of Appointment of a Professor in a Canadian University.
*Courtesy: private source.*

Persons

Author of the act: The University
Author of the document: The University
Addressee: The appointee
Writer: The President of the University

The two documents examined above are expressions of two different juridical systems. This is particularly evident from the answers to the first question. In fact, while a measure of delegation of power exists in every system, the nature of such delegation varies from system to system. In the case of the university, delegation is given the nature of competence by the existence of a written regulation providing various level of hierarchy with the authority and capacity of accomplishing defined actions. In the case of George the Third, while the Executive Council *de facto* exercised the choice of the servants of the Crown, it had no special legal qualifications for that act, and the authority and capacity for it resided in the person of the King. Even the Assembly of Lower Canada could exercise no influence on the nomination of a single servant of the Crown in the colony, and the Lieutenant-Governor was not free to follow his own judgement, being tied down by detailed instructions and constant supervision by the Executive Council, which made him legally incompetent.[11] The nature of delegation at the times of George the Third explains the meaning of the words of Charles II, quoted at the beginning of this chapter: "My actions are my ministers'." The ministers of the King carried out actions for which he was the only competent person, so that those actions were historically theirs but juridically the King's.

Moreover, in the two juridical systems under examination, delegation, beyond having a different nature, proceeds in opposite directions and is linked to responsibility in different ways, as shown by the analysis of the two documents.

---

11. See R. MacGregor Dawson, *The Government of Canada,* fifth edition, revised by Norman Ward (Toronto 1970), pp. 12-13. See also J.R. Mallory, *The Structure of Canadian Government* (Toronto 1984).

As to authorship, it may be worthwhile to restate that the author of the act is the person whose will produces the act. If this person is an abstract entity, like a university, its will coincides with the will of its representative(s) who act(s) in its name. The author of the document is the person having the authority and the capacity, that is, the competence to issue the document. If such competence is exercised in the name of another person, the author of the document is this latter person. The writer of the document is the person competent to articulate the writing. The addressee is the person to whom the document is directed within its intellectual form, and may or may not coincide with the person to whom the document is issued or delivered. For example, a letter of appointment of a professor is directed to the appointee, notwithstanding the fact that we may be examining the copy sent to the department concerned. Here, we are still concerned only with the persons concurring in the creation of a document, not yet with those who may intervene in the document early in its formation process or interact with the document at later stages. The three necessary persons intervening in the creation of a document are to be identified with reference to the original. The persons involved in the formation of drafts and copies may be different.[12] Let us take as example a type of document with which we are all familiar: a set of guidelines on any subject issued by the Association of Canadian Archivists, and addressed to its membership. Examining the original, that is, the version published in the bulletin of the association (the first perfect document, that is, the first document issued in a form which enables it to produce the consequences wanted by the author), we can say that the author is the association, and the writer is the concerned committee. If we look at the draft sent to the executive of the association for approval, it is clear that the author is the committee, and so is the writer. However, an identification of the persons involved in the formation of drafts and copies has no bearing on the definition of the relationships between the document, the act generating it, and the juridical system within which it is created. From a diplomatic point of view, it has relevance only

---

12. For an exposition of the concepts of original, draft, and copy, see pp. 48-53.

at the last stage of diplomatic criticism. Therefore, it will be discussed in the context of the analysis of the extrinsic and intrinsic elements of documentary forms, in the fifth chapter of this book.

Why is it so important not only to determine the persons concurring in the formation of a document, but also to distinguish, among them, who played which role? Certainly, it is important for historical purposes, but beyond them, it is essential for identification and authentication purposes, considering that in the cases of so many documents of past centuries we do not know the provenance, either archival (who created or received them), or custodial (who kept them, and who gave them to the archival institution); neither do we know their documentary context (they are found in groups of documents to which they do not belong). The determination of these persons is also useful to understand and control the many multi-provenance series resulting from frequent administrative changes, and to reconstruct organization, functions, competences, activities, and procedures of records creators, even when we have other sources about them, because documents are the most impartial and authoritative source.[13]

Undoubtedly, archival description benefits the most from the identification of the persons. It is true that archivists rarely describe individual documents, as diplomatists do, but they write administrative histories, and are deeply concerned with authority control. Authority control involves vocabulary control, and David Bearman writes that an "area in which controlled vocabularies may be powerful is the retrieval of names of entities ... that are unambiguously involved in the creation of records." He adds that "correct identification of fields that are candidate for authority control in archives is only the first step." He lists some of those fields as contained in the National Information Systems Task Force data dictionary and the MARC format. It is possible to see a clear correspondence between them and those identified by diplomatics. For example: "creator" can be rendered as "author," "occupation"

---

13. For example, when this writer worked as archivist in the State Archives of Rome, she examined diplomatically the so-called documents of the French Government in Rome (1809-1814), and found out that, contrary to the information provided by statutes, regulations, and the like, there never was an autonomous French government in Rome.

as "qualification of signature," and "relator" as "writer."[14] Of course, the usefulness of diplomatic terminology for purposes of authority control is not limited to the terms already introduced in this book, but may be extended to the major part of the terms used by diplomatics to express its fundamental concepts. The final chapter of this book will deal with this matter.

To return to the persons involved in the creation of a document, it may be observed that such persons have been discussed in relation to textual documents, be they on paper or on machine readable form. With electronic records, identification of the persons may require a more detailed study of procedures and processes, but is based on the same principles. What about aural and visual documents? This writer does not believe that any difficulties are to be encountered in extending those concepts to them. Their application may even be easier than with textual documents, because archivists who deal with special media are accustomed to distinguishing, particularly in visual documents, between the persons competent and responsible for their content, their articulation, their formation, and their form. They even have a special vocabulary distinguishing those persons, and the only thing which remains to be done is to establish the correspondence between the terms of that vocabulary and diplomatic terms.[15]

Identifying the persons concurring in the creation of a document serves many archival and historical purposes, but to diplomatists it was and is essential for one main purpose: to define the nature of documentary forms, their public or private character.

## The Nature of Documentary Forms: Public and Private Documents

The general validity of the distinction between public and private documents is easy to acknowledge, but to find a reliable

---

14. David Bearman, "Authority Control Issues and Prospects," *American Archivist* 52 (Summer 1989), pp. 290-292.

15. It is this writer's hope that this book will generate another book, in which archivists knowledgeable about special media archives will make an effort to apply diplomatic concepts to the material in their care. This book is conceived as a starting point, not as a point of arrival!

criterion for it is another matter. The language of ancient Rome made that distinction apparent: public was what involved the people as a whole, private had the negative meaning of that which was not public, that is, "deprived." Thus, a private document was a document created by a person "deprived" of public office, without public title. Given the circumstances of its creation, a public document dealt with matters concerning all the people, gave full faith and credit to the facts it attested, belonged to the people (it had to be given to the Tabularium, the public archives, as a condition for its effectiveness), and was accessible to the people.[16] Over time, in different juridical systems, the term public came to mean any one or more of the characteristics mentioned above as consequences of public nature, quite independently of the circumstances of creation of the document to which it was attached, so that a document created by a private person could be considered public if it presented any one of those characteristics.

Today, in our juridical system, private has still the negative meaning of non-public, while public lacks a standard definition. In the Anglo-Saxon common-law system, a document may be defined as public with regard to either (1) its *provenance* (if it is created by a public authority, that is, by a body having jurisdiction on matters relating to, or affecting the whole people of a state, nation, or community), or (2) its *pertinence* (if it relates to matters regulated by public law, that is, by the law concerned with the state in its political or sovereign capacity), or (3) its *effects* (if it makes full faith and credit), or (4) its *ownership* (if it belongs to a public body), or (5) its *destination* (if it is accessible to the public).[17] From a legal point of view, if a document is public as to provenance, it is also public as to pertinence, ownership, and effects, but not necessarily as to destination. Only public documents result from the exercise of the state's political, legislative, administrative, judicial, consultative,

---

16. See Luciana Duranti, "The Odyssey of Records Managers," *Records Management Quarterly* 23 (July 1989), pp. 8-9.

17. If we try to define what public nature, public law, and public body mean within that same juridical system, we may get even more confused. Those who are interested in finding out from whence all the confusion derives may read George H. Kendal, *Facts* (Toronto 1980), pp. 43-55.

and controlling functions. However, if a document is private as to provenance, it can be public in other respects.

From an archival point of view, documents are public when they are created or received by a public office, that is, publicity is conferred on documents not only in relation to their provenance but also in respect of the fundamental rule governing every collectivity, according to which each individual element acquires the nature of the whole of which it is part. An archival fonds is a documentary whole; if the fonds is public as to provenance, each document within it is public as well, and vice versa. For example, a letter written by a government department to a private individual is a public document as to provenance, but it is private archivally, because it is part of the archival fonds created by the addressee.

Diplomatists have debated for a long time about the meaning of public document versus private document. Harry Bresslau considers to be public only the documents issued by independent or semi-independent authorities, and restricts the concept of authority only to Emperors, Kings, and Popes, mainly because he deals with medieval documents: we may take his "authorities" to mean government bodies.[18] Artur Giry defines only private documents, considering them to be the documents related to matters of private law and issued by non-public persons.[19] Cesare Paoli defines as public the documents issued by public authorities in public forms, either related to public matters or to specific persons and places, and as private the documents created according to private law and issued by notaries and private persons.[20] Alain de Boüard calls public the documents issued by public authorities. He feels the need, however, to specify that private documents are not only those issued by private persons but also those issued by public persons which are associated for their nature with private law, and similar in their forms to those created by private persons.[21] Finally, Georges Tessier accepts the general opinion that public documents are those issued by public authorities, but agrees with Giry and Paoli

---

18. Harry Bresslau, *Handbuch der Urkundenlehre für Deutschland und Italien* 2 vols. (vol.1: Berlin, 1889; vol.2: Leipzig, 1912-1931), vol. 1, p. 3.

19. Artur Giry, *Manuel de diplomatique* (Paris, 1893), p. 823.

20. Cesare Paoli, *Diplomatica* (Firenze, 1899), p. 27.

21. Alain de Boüard, *Diplomatique générale*, pp. 40-41.

in that private documents are those issued by private persons and related to matters which are the concern of private law.[22]

It is clear that there are three elements identified by different diplomatists as central to the public or private nature of a document:

1. the person creating the document (provenance)
2. the juridical content of the document (pertinence)
3. the forms of the document.

There is no doubt that these elements are all substantial to the definition of the nature of a document, but why are we defining that nature? There is no such thing as an absolute definition. We always define for a purpose, and a definition must be tailored to that purpose. There may be many correct substantial definitions of the same thing, but only one appropriate for each given point of view. Diplomatics defines the public or private nature of a document for purposes of identification and evaluation. The means to reach these purposes is the analysis of the forms of the document. Such analysis needs to be based on an hypothesis in order to be fruitful, and the hypothesis advanced by diplomatics is that the forms of a document are conditioned by its author. Therefore, it is vital for diplomatics to define the nature of a document in relation to its author and to make of that definition the fundamental assumption around which the entire diplomatic criticism revolves. The forms of the document cannot constitute the core of the definition. They are not an antonomous entity (the concepts of public and private form have no theoretical existence[23]) and they are the object of diplomatic enquiry. Since they are what has to be analysed, identified and evaluated, they cannot be defined by means of terms and elements which have to be identified and evaluated on the basis of the definition of that entity.

As to the juridical content of the document, the most valid argument against its adoption as the focus of the definition of

22. Georges Tessier, *La Diplomatique* (Paris 1966), pp. 65, 99.

23. This writer is referring to theory in its meaning of verifiable generalization of a high order which is derived from the observation of phenomena and explains them.

public and private documents is advanced by Alessandro Pratesi. He states that, in addition to the facts that there is no clear distinction between public and private law in any juridical system and that a single document may deal with matters concerning both, diplomatic analysis is not applied to the content of individual documents, but to their formal elements; therefore content should not be the foundation of its theoretical construct and methodology of enquiry.[24]

The conclusion of this discussion is that the definition of the nature of a document which is most suitable to diplomatic purposes must put the document into relationship with its author. Accordingly, *a document is public if it is created by a public person or by his command or in his name*, that is, if the will determining the creation of the document is public in nature. A public person is a juridical person performing functions considered to be public by the juridical system in which the person acts and in so doing, vested with the exercise of some sovereign power. By contrast, *a document is private if it is created by a private person or by his command or in his name*; that is, by a person performing functions considered to be private by the juridical system in which the person acts. This implies that the documents created by a public person in his private capacity, that is, performing private functions, are private in nature. Here, it is essential to reiterate that the creator referred to is the author of the original document. In fact, some confusion may derive from the status of transmission of the document when trying to establish its public public or private nature. For example, if a public officer copies public documents for his own private purposes, and includes them within his own fonds, there could be the temptation to consider those copies to be private documents. On the contrary, they are public documents for two fundamental reasons, the first being of diplomatic nature, and the second of both legal an diplomatic nature. It has been stated that a document is public if its author is public, the author of a document being the person whose will determines the creation of the complete document (medium, physical and intellectual form, content). A person who copies a document has an influence on its medium, sometimes

---

24. Pratesi, *Elementi di diplomatica*, p. 24.

on its physical form (eg. imitative copy), but it is not his will generating the whole document. Moreover, within most juridical systems, the legal system expressly considers to be public as to provenance and ownership the informational content of documents created within the public sphere, at the point of prosecuting those who disseminate it without authorization. And, diplomatically, documents must be identified and evaluated within their juridical system. Thus, once again, the will is the prevalent element which identifies both the author of a document and its nature.

It seems that at this point we have solved all the problems inherent in the definition of the nature of documents and may proceed to the examination of the two categories of public documents and private documents. Why, then, did such an obvious conclusion not find consensus among diplomatists: what obscured their vision and confused their ideas? The answer is easy: too many of the documents they were examining could not be categorized so simply. To say that a document owes its nature to the nature of the will creating it is easy, but what is the nature of a document generated by more than one will when those wills are of a different nature? We do not need to look back to medieval documents to see examples of this situation. We have many such documents all around us, maybe on our desk. Think of an income tax return, a census form, an information and complaint for a crime, or a contract between a public and a private person. Diplomatists approached the problem either by broadening and specifying the definitions of public and private documents to include all observed cases, or by creating a third category of documents. We have already examined the attempts made to broaden the definitions, and discussed the reasons why they are unsatisfactory. Diplomatists of the Italian school, in particular Alessandro Pratesi, see the need to identify a third category, "semi-public documents," defined as the documents issued by private persons by command of and in the forms established by public persons.[25] A number of arguments may be used against this proposal. First, if the will originating the document belongs to a public person, and the private person has no choice but to obey, the author is definitely the public person, and

---

25. Ibid., p. 25.

the private person is the writer (e.g., an income tax return). Secondly, if a meeting of wills of a different nature is a necessary condition of the existence of the document, the procedures and forms imposed by the public person are necessary as well, and because diplomatics is mainly concerned with procedures and forms, it is possible to consider the will determining them as the prevalent will in the document. Thirdly, if we accept the third category identified by Pratesi as distinct from the other two, we have to contemplate the possibility of constituting a fourth category for all those documents which are created by a convergence of wills of opposite nature in a form typical of private acts (e.g., a deed of land between a public and a private person). How do we consider this type of document? David Bearman proposes two alternative approaches: (1) "one could arbitrarily take the position (sometimes too true) that such a contract is an uneven match and that the private citizen is in effect always the second party," or (2) one could simply follow the convention already established by diplomatics that, in a contract of reciprocal obligation, the first party is the author and the second party is the addressee.[26] This writer thinks that the second approach is the most appropriate because it avoids the creation of another category by assimilating this type of document with all documents of reciprocal obligation, concentrates the analysis on procedures and forms rather than on philosophical conceptions which may differ within the same juridical system, and is therefore consistent with diplomatic theory. In conclusion, she believes that (1) documents are either public or private depending on the nature of their author; (2) the type of document identified by Pratesi as semi-public belongs to the category of public documents; and (3) the documents resulting from reciprocal obligations between private and public persons, if not created according to procedures and forms imposed by the public person, may be either public or private, depending on the nature of the first party. It remains established that the status of transmission has no influence on the public of private nature of a document.

The preceding exposition and discussion has been conducted from a diplomatic point of view, and serves the diplomatic pur-

---

26. David Bearman, letter to the writer, 28 October 1989.

poses of identification and evaluation of individual documents. How can the diplomatic theory about public and private documents be applied to groups of documents strictly interrelated by the act producing them? Most modern bureaucratic acts are compound acts, and specifically acts on procedure. These acts do not manifest themselves in one document, but in files and even in series.[27] By answering the above question, diplomatic theory could be extended to meet the boundaries of archival theory. David Bearman writes:

> The archivist adopts a view entirely with respect to *transmission* of the document—it is public if it is created by or sent to a public institution. Could we find in the principle of the "will" that commands the creation of the document a similar principle? Obviously, it helps us to distinguish matters of procedure from those of process within the body of records created in a public agency. I think that if we view the documents created to participate in juridically relevant procedures as being created in a form commanded by the public authority responsible for the procedure, we can distinguish between the large number of documents received by public agencies those which could not be considered records (e.g., they were not intended to participate in an established procedure) and those which are records. Of the latter, many have also their form prescribed by the will which authors the procedure.[28]

What Bearman is suggesting as a means for broadening the application of diplomatic theory and reconciling it with archival theory is to consider a fourth element as the possible focus in the definition of public and private nature, namely *the author of the procedure*.[29] Because any development of diplomatic theory should not contravene its own fundamental principles, Bearman's proposal should be examined in the same light as the other proposed elements. First, is the identification of the author of the procedure which generates the documentation of an act as the element deter-

---

27. Pp. 76-77.
28. Bearman, letter to the writer, 28 October 1989.
29. The other three were:(1) the person creating the document; (2) the juridical content of the document; (3) the forms of the documents.

mining public or private nature of the resulting documents in contrast with the fundamental assumption that the forms of a document are conditioned by its author? The answer is clearly negative, because forms are conditioned by procedure as well. This latter relationship has no influence on the former and is independent of it. Moreover, procedures and their authors are a concern of diplomatics without being the direct object of diplomatic analysis like forms. In fact, procedures are revealed by the examination of forms, not by direct observation of the procedures themselves. Secondly, we should examine the extent of the usefulness of Bearman's criterion. Is there a clear distinction between public and private procedures in any given juridical system, and, in the presence of a group of documents resulting from an interplay of private and public procedures, is it possible to determine which one is prevalent, or, if they are equal, which one belongs to a first party? The answer to these questions is positive because the concepts of public and private procedure are theoretically definable in relation to their author, and differ from the concepts of public and private forms, which are not necessarily linked to the nature of those creating them.

On the basis of this brief analysis, it seems perfectly adequate in diplomatic terms to consider procedure as a valid focus in defining public and private documents in all those cases in which an action results in a body of documents, rather than in one isolated document. However, when examining single documents, particularly for purposes of identification, even if they belong in a group, it is essential to establish their nature, public or private, in relation to the nature of their author. Diplomatics should broaden its scope by developing new principles and methods for the analysis of documentary material not created at the time in which its theory was formulated, but it should not change the foundations of that theory!

A broadening and enrichment of diplomatic theory and methodology is especially needed for the analysis of "the genesis of public and private documents". This will be the purpose of the fourth chapter of this book, which will deal with procedures and their authors.

# Chapter 4

# The Procedure of Creation of Documents

The heavens themselves, the planets and this centre
Observe degree, priority, and place,
Insisture, course, proportion, season, form,
Office and custom, in all line of order.

Shakespeare, *Troilus and Cressida*.
Act i, sc. 3, l. 85

Systems owe their integrity to their logical cohesion, that is, to
the consistency of their elements with their purpose and with each
other; to the existence of distinct boundaries between those elements; and to the definition of an internal order. A system is made
of building blocks and of a purpose, which rules it from the outside,
determining the boundaries in which the system is designed to
operate. The ultimate aim of a system is to provide security amid
change, and a force for its own continuity.

Diplomatics saw the documentary world as a system, and built
a system to understand and explain it. Early diplomatists rationalized, formalized and universalized document-creation by identifying within it the relevant elements, extending their relevance in
time and space, eliminating the particularities, and relating the
elements to each other and to their ultimate purpose. The identified
elements were the juridical system, which constitutes the necessary
context of document-creation; the act, which is its determinant
cause; the persons, who are its agents and factors; the procedures,
which guide its course; and the documentary form, which allows

document-creation to achieve its purpose by embracing all the relevant elements and showing their relationships. These elements are building blocks which have an inherent order: in fact, they can be analysed in a sequence from the general to the specific, following a natural method of inquiry. However, such a method can be adopted only when the reality is fully observable or attainable. If this is not the case, a knowledge of the abstract characteristics of the system and its component parts, and of their relationships, makes it possible to understand the essential aspects. By referring to this knowledge, each single element of the system can be used as a key to all the others, and can lead to the comprehension of the greater whole. This is the analytical method of inquiry, which is applied by the so-called "exact sciences" and which, in a process of discovery, tends to precede the method of moving from the general to the specific, and allows the formulation of generalizations.

The diplomatic process of abstraction and systematization decontextualized the elements of document-creation, and made explicit what was implicit, so that contradictions could be recognized and relationships understood. This loss of context through generalization did not undermine the validity of the results. In fact, if it is true that familiarity with context is characteristic of human life, it must also be accepted that something which has become familiar can be recognized and understood in a different context, and can serve as a reference point from which the relevance of changes in the context can be measured.

The second and third chapters of this book presented the juridical system, the acts and the persons, following the logical progression which underlies the quest of the archivist. This chapter will proceed one step further in that sequence by presenting the procedures which, within a juridical system, are followed by the persons in order to accomplish acts resulting in documents. That is, this chapter will discuss the *genesis of documents*.

The genesis of documents is an elaboration of routines. Routines have a built-in resilience, which enables them to absorb changes occurring between one set of routine actions and its repetition. Once the steps involved in the performance of an action are established or ascertained, decisions or enquiries are no longer needed. Confronted with a problem, all one has to do is to link it to the familiar, and by analogy what is relevant can be easily found. The genesis of documents was seen as a sequence of two sets of routines, which were called by early diplomatists *actio* and *conscriptio*, and by this writer *moment of action* and *moment of documentation*.[1] "But general diplomatics does not stop at this first result"—writes Alain de Boüard—"it composes, with the whole of the concrete categories, an ideal act. By doing so, diplomatics allows the analytical examination to put into evidence and to study in their logical sequence all the facts which can determine the creation [of a document] or concur in its formation."[2] General diplomatics considers the moment of action and the moment of documentation to be in their essence two procedures, which may develop either in parallel or in sequence, and identifies the steps involved in each of them.

## The Moment of Action and the Moment of Documentation: Two Procedures

A *procedure* is the formal sequence of steps, stages or phases whereby a transaction is carried out. Whether it is regularized in written rules or some other means, every procedure tends to have a structure. On the basis of this fundamental assumption, diplomatists of medieval documents identified the typical components of the two procedures guiding action and documentation which were evident in the formal elements of the archival material available to them: documents issued by public authorities, and notarial deeds.[3]

The examination of documents issued by public authorities reveals the existence of two distinct types of actions, or acts: those which were undertaken by the authority on its own direct initiative,

---

1. Alain de Boüard, *Diplomatique générale* (Paris: Editions Auguste Piscard, 1929), p. 62. De Boüard mentions the fact that medieval formularies distinguished between the *tempus in quo ea facta sunt superquibus litera datur* (the moment when the facts about which documents are written take place), and the *tempus in quo datur litera* (the moment when documents are written). He also points out that Mabillon separated in his examination the *res transacta* (transaction) from the *instrumentum confectum* (recording) (p. 60).

2. "Mais la diplomatique générale ne s'en tien pas à ce premier résultat. Elle compose, de l'ensemble des catégories concrètes, un acte idéal permettant à l'analyse d'exposer et d'examiner dans l'ordre de leur succession logique tous les faits qui ... purent provoquer la naissance [du document] ou concourir a sa formation." De Boüard, p. 66. This and all the following quotations from French texts are translated into English by this writer.

3. For an illustration of the concept of public authority see pp. 98-99.

of its own will, in the context of its political-sovereign capacity; and those which were initiated by other juridical or physical persons, public or private. In the former case, the moment of the action comprises one *simple act* consisting of the order given by the authority to its chancery to compile the document expressing its will.[4] This act is called *iussio* (command), and does not have an evident procedural nature.[5] In the latter case, we have a *compound act on procedure*, consisting of well-defined stages or phases.[6] The first phase is called *petitio* (petition). The petition is the request of a physical or juridical person to the authority to accomplish an act. Petitions were customarily presented in writing, in a predetermined form, to the chancery of the authority. Sometimes, the petitioners were given a hearing to express their requests, but, from the thirteenth century, petitions tended to be made only in writing. The second phase is called *intercessio* (intercession), and consists of the propitiatory intervention of persons close to the authority. The intercession was rarely presented in person by its author; it used to take the form of a letter of recommendation or reference providing information on the petitioner and expressing support for the petition. The third phase is the *interventio* (intervention). The content of this phase changed through the centuries as the juridical system evolved. In the early Middle Ages, the *intervenientes* (those who intervened in the transaction) were persons who could be damaged by the transaction, and, by their presence, guaranteed the validity of the act. Later, with the weakening of the sovereign power, the intervening persons were the magnates who gave their consent to the transaction. Between the tenth and twelfth centuries, the intervening persons became simple witnesses to the action, because the authority did not ask for their opinion or consent any more. Some-

---

4. We have a *simple act* when the power of accomplishing the act is concentrated in one individual or organ (p. 75).

5. A simple act, and, in the specific case, the *iussio*, has however a hidden procedural nature, which will be discussed later on in this chapter.

6. A *compound act on procedure* is an act which derives from an established sequence of different acts having the common aim of making possible the accomplishment of the final act (p. 76).

times, they did not even witness the act, but were mentioned in the document to give it solemnity. The fourth phase was the *iussio*, the order given by the authority to the chancery to compile the document embodying the transaction. This phase exists also in those actions which were routinely accomplished by the chancery without the knowledge of the authority. In this case the *iussio* is implicit in the regulation for the functioning of the office.

The diplomatic examination of medieval notarial deeds reveals the absence of an evident procedure prior to the compilation of the document representing the transaction. The *rogatio*, that is, the request to compile the document, presented orally by the parties to the notary, does in fact correspond to the *iussio* expressed by public authorities, even if it has the diplomatic configuration of a *contract*.[7]

The analysis of medieval documents shows that the set of routines, or procedures governing the moment of the action comprised a minimum of one to a maximum of four phases, depending on who takes the initiative for the transaction, whether its author(s) or somebody else. This analysis is not completely convincing. If one extracts the relevant facts from their historical and documentary context, and avoids considering every action as necessarily endowed with a definite form,[8] one can clearly see that every transaction begins with an *initiative* and manifests itself by means of a *deliberation*. In deed, a transaction differs from any other fact because it is prompted by an act of will aimed to produce consequences, that is, to create, maintain, modify or extinguish situations. This also implies that a transaction derives not only from an initiative, but also from an assessment of the situation that it intends to influence. Such an assessment necessarily follows the collection of relevant information and the analysis of the data assembled. Thus, it is possible to identify two other phases between the initiative and the deliberation, phases which might be called *inquiry* and *consultation*.

---

7. We have a contract when the power of accomplishing the act belongs to two or more interacting parties (p. 76).

8. Jack Goody calls this tendency "actional formalism." His arguments, even if only indirectly related to the subject of this chapter, may be of great interest to archivists. (Jack Goody, *The Logic of Writing and the Organization of Society* (Cambridge: Cambridge University Press, 1986), p. 144.)

To sum up, if we consider the procedures governing the moment of the action to be a logical system rather than a set of formal manifestations, we can say that every such procedure, quite independently of its author(s) and its initiator(s), comprises four phases: *initiative, inquiry, consultation* and *deliberation*. The correspondence between these phases and the *petitio, intercessio, interventio* and *iussio* is obvious, but while these latter phases are unequivocally linked to a specific historical and documentary context, those proposed by this author are "decontextualized" and are therefore recognizable in every context, even when they do not materialize in visible actions or in documents.

The procedure governing the moment of documentation as seen by diplomatists of medieval documents was formalized in office routine. In the chanceries of public authorities, this procedure began with the *compilation of the draft* of the document, which was followed by the *preparation of the fair copy*. Those chanceries which made use of registries *transcribed the document*, entirely or partially, either before its validation, that is, as the third phase of the procedure, or after validation, as the fifth phase. The most solemn phase was the *roboratio*, or validation of the document, which was made according to different systems, the most common being (a) the intervention of the author, who either subscribed or put a particular sign; (b) the intervention of the highest official of the chancery, whose subscription attested that the document corresponded to the will of the authority; (c) the intervention of witnesses, which gave solemnity to the document; (d) the drawing of special signs (monograms, *rota*, etc.); and (e) the affixing or appending of the seals. The following phase, not always present, consisted of the *computation of the tax* to be paid by the addressee, and the writing of its amount on the margin of the document. The final phase was the *delivery* of the document, or its *publication*, if the general public had to be notified of its content.

The creation of notarial deeds followed a very similar procedure. The first phase was the *compilation of the draft*. It used to take place in two stages: at the moment of the request by the parties, the notary wrote the essential data on the *verso* of the parchment destined to contain the document (i.e., names, action, description of the property, etc.); later, the notary compiled a fuller draft omitting only the formulas which were identical in all analogous deeds. The second phase was the *preparation of the fair copy* in its definitive form, that is, with inclusion of formulas. When the authority of notaries was fully established (after the tenth century), this phase disappeared, because the preservation by the notary of the draft of the deed was considered sufficient evidence of the existence of the transaction. The third phase was the *subscription*. Usually the parties did not subscribe, while we often find the subscription of witnesses or their signs. When the notary became a public official, notarial deeds started to present only the subscription of the notary. Its function corresponded to the validation in public documents. The final phase was the *traditio*, that is, the delivery of the document to the concerned party.[9]

From the analysis of medieval documents, it is possible to conclude that the set of routines, or procedures, governing the moment of documentation comprises four necessary and two possible phases, as follows:

| | |
|---|---|
| 1) compilation of the draft | (necessary) |
| 2) preparation of the fair copy | (necessary) |
| 3) registration | (possible) |
| 4) validation | (necessary) |
| 5) computation of tax | (possible) |
| 6) delivery | (necessary) |

Like the result of the analysis of the procedure governing the moment of action, the above schema is not convincing, and not only because it is not transportable to a different historical and documentary context. It is unsatisfactory also within the context under consideration. In fact, it is valid only for documents of an external and operational character, and for contracts. To make this schematization valid for all the documents created by an office, it is probably more appropriate to consider all of the six phases as possible, and each of them as a possible *compound act*: in fact, the

9. The sets of routines governing the moment of the action and the moment of documentation as described by diplomatists of medieval documents can be found in all major manuals of diplomatics. This author has followed in particular the schematization presented by Alessandro Pratesi, *Elementi di Diplomatica Generale* (Bari: Adriatica Editrice, n.d.), pp. 29-51, and by de Boüard, *Diplomatique*, pp. 62-111.

validation of a document, for example, may be an *act on procedure*, and the preparation of a draft a *continuative act*.[10]

## The Moment of Action and the Moment of Documentation: One Integrated Procedure

While the medieval document concentrates the information on the event of which it is the instrument and the product, and represents a kind of knot of information, the contemporary administrative document is only one of the elements of atomized information. The piece of a dossier has interest only if it is at its place within the dossier, which is itself the basic unit, the basic instrument of administrative activity.[11]

Gérard and Christiane Naud have pointed to the most obvious fact which differentiates the genesis of medieval documents from that of modern documents. Each medieval document contained the whole transaction generating it, and its creation, as the apex of the transaction, was either sequential to it (probative documents), or parallel (dispositive documents),[12] that is, perfectly distinguishable from the transaction as an expression of will. On the contrary, each modern document incorporates only one phase of the transaction, or even less, and its creation, as a means of carrying out the transaction, is integrated in each of the phases through which the transaction develops, and is not distinguishable from the action of the will. This fact invalidates the definition of the moment of the action and the moment of documentation as two separate sets of routines, or two distinct procedures. They are still two *conceptually*

---

10. The concepts of compound act and act on procedure are explained in note 6. A *continuative act* is a compound act made of a sequence of identical acts accomplished by the same individual or organ (p. 76).

11. "Alors que la charte médiévale concentre l'information sur l'histoire dont elle est l'outil et le produit, représentant une sorte de noeud d'information, le papier administratif contemporain ne livre lui, que l'un des éléments d'une information atomisée. La pièce d'un dossier n'a d'intérêt que si elle est à sa place dans le dossier, qui est, lui, l'unité de base, l'outil de base, du travail administratif." Gérard et Christiane Naud, "L'analyse des archives administratives contemporaines," *Gazette des Archives* 115 (4e trimestre 1981), p. 218.

12. *Dispositive* documents are those executing an act. *Probative* documents are those providing evidence of an act which was executed before being documented (pp. 65-69).

*distinct moments,* nevertheless, even if they are considered integral parts of one procedure. This can be demonstrated by analysing the ideal structure of the integrated procedure which generates documents. Such a structure, independently of historical-administrative context, author and purposes, comprises two or more of the following phases:

1) Introductory phase or *initiative.* It is constituted by those acts, written and/or oral, which start the mechanism of the procedure. Examples of documents created in this phase are petitions, applications, claims, drafts of bills.[13]

2) Preliminary phase or *inquiry.* It is constituted by the collection of the elements necessary to evaluate the situation. Examples of documents created in this phase are surveys, estimates, curricula, technical reports, reference letters.

3) Consultative phase or *consultation.* It is constituted by the collection of opinions and advice after all the relevant data have been assembled. Examples of documents created in this phase are agendas, minutes, memoranda, discussion papers.

4) Deliberative phase or *deliberation.* It is constituted by the final decision-making. Examples of documents created in this phase are drafts of appointment notices, contracts, laws.

5) Controlling phase or *deliberation control.* It is constituted by the control exercised by a physical or juridical person different from the author of the document embodying the transaction, on the substance of the deliberation and/or on its forms. Sometimes, some form of control is necessary to insure the effectiveness of the deliberation and its enforceability. Examples of documents created in this phase are letters of transmis-

---

13. Because these phases are described out of context and with no reference to the author of the procedure and/or or the documents listed as examples, the term "created" is used in the general sense. Also, it has to be pointed out that, from a logical point of view, all these phases are included in every procedure, even if not in a formal way.

sion, memoranda, and definitive compilations of the documents embodying the transactions.

6) Executive phase or *execution*. It is constituted by all the actions which give formal character to the transaction (i.e., validation, communication, notification, publication). The documents created in this phase are the originals of those embodying the transactions and, for example, registrations, letters of transmission to a printing shop, or to a newspaper.[14]

Some procedures are very formalized: each phase is distinct from the others and easily recognizable. Other procedures are very informal, and some phases take place at the same time or do not leave a documentary residue. Nonetheless, every transaction passes through the above procedural phases, which constitute a closed logical system.

The schematization presented shows an integrated procedure, each phase of which comprises both the moment of action and the moment of documentation. However, while in the first three phases all the documents created are interlocutory with respect to the transaction as a whole, that is, they are necessary either to initiate or to develop the transaction, but are not the ultimate purpose of the procedure and result of the transaction in the subsequent three phases, the focus of each action is the preparation, completion and perfecting of the documents embodying the transaction. This implies that within each of the first three phases the moment of documentation, as well as the moment of action, reaches its completion, while within each of the latter phases only the moment of action does—because, with respect to the document(s) embodying the transaction, the moment of documentation starts on the fourth phase with the compilation of one or more drafts; proceeds in the fifth with the preparation of the fair copy following the control exercised on the substance, articulation and mode of formation of

---

14. Paola Carucci, *Il Documento Contemporaneo. Diplomatica e Criteri di Edizione* (Roma: La Nuova Italia Scientifica, 1987), pp. 47-63. Carucci, illustrating the procedure typical of contemporary Italian administration, does in fact describe the same procedure that this writer has rationalized.

the draft; and ends in the sixth with the creation of the original document(s) by means of validation and/or publication, eventual registration and delivery and/or inclusion in the file.

An example may clarify this point. Let us examine a very typical procedure, the appointment of a university professor. The original document which will embody the transaction is the letter of appointment sent by the president of the university to the appointee. Thus, the procedure which will be described refers to the genesis of that specific document, even if its documentary residue will accumulate in files, the closing document of which will ideally be a copy of the letter of appointment, even when followed by a copy of the appointment notice sent out for publication. The complete file, containing all the documents produced during the procedure, in the form in which they participated in it (that is, in draft form if they participated in the procedure in that form; in original if they did so, etc.), will be only in the office of the head of the department concerned. Duplicates of the file, either partial or complete, will probably be among the records of the members of the search committee, in the dean's office, in the president's office, among the Board of Governors records, and in the appointee's fonds. Traces of the transaction may also be found in the immigration office. This procedure is very formal and strictly follows the six ideal phases:

1) *initiative* Issuing of an advertisement for the position.

2) *inquiry* Collection of applications, curricula vitae, reference letters, copies of publications; interviews.

3) *consultation* Discussion of the data assembled by the members of the search committee. Recommendation to the department head.

4) *deliberation* Offer of the position to the applicant and receipt of his/her acceptance. Compilation of a form by the department head, with inclusion of the relevant data of the transaction (this is very similar to the preparation of the first draft by the medieval notary).

5) *deliberation control* Control of the data included in the form as to their substance, by the dean; as to their completeness and appropriateness, by the president and the Board of Governors. Eventual correction. Approval of the definitive document.

6) *execution* Issuing of the letter of appointment by the president to the appointee. An appointment notice is issued by the head of the department and sent for publication.

From this example it is clear that the documentation moment of the last three phases can be considered as a *continuum*, the purpose of which is the creation of one perfect, enforceable document embodying the whole transaction. On the contrary, each of the first three phases comprises one or more integrated and complete procedures aiming to facilitate a transaction through creation of the documents typical of that phase. Such documents, while interlocutory with respect to the main procedure, are final with respect to the subordinate procedures. This can be demonstrated by analysing the *initiative* phase of the procedure leading to the creation of the letter of appointment: "Issuing of an advertisement for the position." The document the genesis of which we are going to examine is the advertisement. Relative to it, the following procedure is a complete transaction:

1) *initiative* Identification of the need for a new position and presentation of a request for it, usually by the head of the department to the competent dean, orally and in writing (memorandum).

2) *inquiry* Collection of data on the financial situation; and on the availability of qualified persons.

3) *consultation* Discussion of the assembled data in order to decide on the content of the advertisement; consultation on the composition of the search committee.

4) *deliberation* Composition of the advertisement in draft form by the search committee.

5) *deliberation control* Control by the head of the department and the dean of the substance of the document (description of responsibilities, qualifications, salary); control by the office of the president of the formulation of the qualifications, so that equity of employment is respected; control by the immigration office of the presence of the prescribed formula giving precedence to citizens and landed immigrants. Compilation of the fair copy.

6) *execution* Printing and distribution of the advertisement for publication. Communication of the document to all parties who might be interested.

This type of analysis could be conducted on any of the phases of the procedure leading to the creation of the letter of appointment, and on any of the procedural phases leading to the creation of the advertisement. In fact, the dissection of the first procedure considered may continue until the genesis of all single documents participating in the main procedure has been examined. However, at one step down in our operation we would already encounter some difficulties, clearly identified by Gérard and Christiane Naud: "The administrative action proceeds by cascades and ramifications, from the general to the specific and vice-versa. The handling of a transaction follows at one time several channels which separate and later rejoin, each service, office or official being entrusted with a part of the total procedure ... the problem is what point of view to adopt, and we think that the archivist must adopt the point of view of the administration which created the archival material."[15] This means that, not only would it be impossible for an archivist to follow all the ramifications of each single transaction, but, more importantly, it would be useless. That kind of operation does not even belong in the work of the special diplomatist, who has the specific purpose of identifying the "typical" transactions of a given administration and describing their ideal structure and interrelationships, so that the entire functioning of the administration can be made evident.

---

15. "L'action administrative procède en effect par cascades et par ramifications, du général au particulier et vice-versa. Le reglèment d'une affaire suit simultanément plusieurs canaux qui se séparent puis se rejoignent, chaque service,

Rather, the archivist needs to distinguish "the stages of an action or the phases of a procedure, because the form of the documents one encounters results from the status of development of the procedure,"[16] from the point of view of the documentary body with which he/she is dealing. Of course, one can object that it is not always necessary to identify forms of documents, particularly considering that we modern archivists do not deal with single documents. Gérard and Christiane Naud directly address this point in a very effective fashion:

> "And if one takes into consideration the fact that a *dossier* rarely coincides with a *file*, being generally smaller or larger, one can see that the unit to be described will necessarily be the dossier or a part of it. If it is necessary to describe a part of a dossier, one needs to point out the element of the procedure from which it results. It is for this reason that we introduce in the description of the content of the files an element that identifies the *action* which results in the existence of the dossier(s) or of the sub-dossiers that it contains. It is for this same reason that we have to find a solution that allows us to place the action from which the described unit results in the context of a more general action, that is, of the mission or characteristic in virtue of which the transferring administration acted."[17]

Thus, an understanding of the procedure governing the genesis of documents is essential to carrying out the descriptive function, but such an understanding can only begin once the form of the document(s) embodying the transaction which make up the dossier has been identified. For example, a university archivist acquires the fonds of a faculty member and encounters a file (Naud's *article*)

---

bureau ou fonctionnaire étant chargé d'une partie de la procédure totale ... leproblème est celui du point de vue duquel se placer et nous pensons que l'archiviste doit adopter le point de vue ... de l'administration productrice des archives." Naud, "L'analyse des archives ...," pp. 218 and 223.

16. "Des étapes d'une action ou des stades d'une procédure, la forme des documents rencontrés résultant en fait de l'état d'avancement de la procédure": ibid., p. 226.

17. "Et si l'on tient compte de ce que le *dossier* coincide rarement avec l'*article*, étant généralement plus petit ou plus grand, on voit que l'unité à décrire sera nécessairement le dossier ou la partie de dossier. S'il faut décrire la partie de dossier, il faut préciser l'élément de procédure dont elle résulte. C'est pour cela que nous introduisons dans la description du contenu des articles un élément identifiant

containing the letter of appointment received by the professor. In order to arrange and describe that file, he/she has to establish first whether the file corresponds to one transaction (Naud's *dossier*), to more than one transaction, or to a part of a transaction. In fact, the file might contain only the material related to the appointment of the faculty member; all the material related to his/her employment relationship with the university; or part of it. This is easy to ascertain if the archivist is familiar with the procedures of appointment and tenure at the university, and with the way in which a faculty member participates in them. If the file coincides with the appointment transaction, it might contain

1) *initiative* Copy of the advertisement for the position; copy of the application and its enclosures.

2) *inquiry* Material related to the department in question, the faculty, the campus, the city. Correspondence aimed at collecting data useful for the interview.

3) *consultation* Correspondence with the head of the department, other faculties, family, etc., on conditions of appointment, of relocation, etc.

4) *deliberation* Original of the letter offering the position; draft and copy of the letter of acceptance of the position.

5) *deliberation control* This phase may produce two different sets of documents in the professor's file: a) if the control is that exercised by the professor over the conditions of appointment as expressed in the letter offering the position, it may or may

---

*l'action* dont résulte l'existence des dossiers ou du dossier ou des sous-dossiers qu'il contien. C'est également pour cela qu'une solution doit être trouvée qui permette de situer l'action dont résulte l'unité décrite, dans le cadre d'une action plus générale, c'est-à-dire de la mission ou de l'attribution en vertu de laquelle l'administration versante a agi" : ibid., p. 218. For those who are not familiar with French archival terminology, it should be pointed out that a *dossier* consists of the documentary residue of an entire transaction, while the concept of *article* corresponds to the concept of file. Therefore, it is quite clear that a file may be a part of a transaction, or may contain a number of transactions (i.e., related to the same matter).

not produce further correspondence, and would take place at the same time as the deliberation phase; b) if the control is that exercised by the department over the qualifications of the professor, the file of the professor may contain copies of the documents supporting those qualifications as provided to the department.

6) *execution* Original of the letter of appointment; copy of the appointment notice.

The example presented refers to a very simple situation, but the analysis develops along lines independent of the complexity of the documentary body under examination. The focus is always the transaction and its procedure, and the starting point is constituted by the documentary forms embodying them. A consideration of the subject—what any group of documents is about—accompanies the analysis; it does not guide it. If different files in the same fonds receive a common description, it is because they either result from consecutive phases of the same transaction; from similar transactions related to different subjects; or from consecutive transactions related to the same subject. Thus, within the same fonds, arrangement and description have to concentrate on the transactions. Besides, if relationships among dossiers included in different fonds can be established on the basis of their common subject, the difference among those dossiers results from the different ways in which their creators have intervened on that subject.[18] For example, a bridge is built or maintained, and the files which result from those two actions are very different. The file produced by the construction operation is voluminous; contains many transactions of different types (administrative, financial, technical); and spans a limited

---

18. Ibid., p. 226. Gérard and Christiane Naud write that, when transferred from the creating office to the archival repository, each dossier should be described by mentioning the following elements: the transferring body, the agent of the administrative action of which the dossier was the instrument, the action, the subject of the action, date and place, elements of form. They also explain with an abundance of examples how to identify action and subject without confusing the two, particularly when the subject is an action itself (i.e., when the action of individuals is the subject of an action of an administration). Ibid., pp. 220-25.

number of years (from the date of the first project to the end of the period allowed for claims related to the actual construction). On the contrary, maintenance is a continuing action, giving origin to files the opening and closure of which depend on the practice of the office. The documents included in this type of file tend to be repetitive as to content and standard as to form. From an archival point of view, the value of the latter files will be different from the value of the former files.[19] This example does more than just demonstrate that the difference between files lies primarily in the action. It shows that the identification of the action and of the procedure guiding it is important not only for arrangement and description, but also for appraisal. It might be observed that, given the bulk of contemporary administrative documents, we neither appraise nor describe file by file. We tend to conduct those operations on larger units, namely series. However, in order to understand and evaluate the content of a series, we examine samples of its component files; then extrapolate the result of that observation to the whole series; and finally describe and evaluate the series with respect to both its components and the whole administrative action of which it is a residue. This need to generalize from the circumstances we observe introduces another aspect of the system built by general diplomatics for understanding document-creation: the categorization of procedures.

## The Categorization of Procedures

When the administrative procedures and the techniques of handling information in the offices evolve rapidly, we constantly encounter the problem of knowing whether this or that concrete category of documents is equivalent to this or that other ancient category ....[20]

---

19. Ibid., p. 222.
20. "Alors que les procédures administratives évoluent rapidement, en même temps que les techniques de traitement de l'information dans les bureaux, nous nous heurtons sans cesse au problème de savoir si telle ou telle catégorie actuelle de documents est équivalente a telle ou telle autre ancienne catégorie ....," ibid., p. 216.

It has been shown how each single procedure presents the same ideal structure, independently of its context, author and purpose. However, the activities involved in carrying out each phase of a procedure vary according to the purpose of the procedure, and so do the documents resulting from those activities. In order to identify and evaluate the activities and their documentary residue, diplomatics has distinguished all possible procedures in four categories, on the basis of their general purpose:

1) *organizational procedures*: those aimed at the establishment of organizational structure and internal procedures, and their maintenance, modification or extinction.

2) *instrumental procedures*: those connected to the expression of opinions or advice.

3) *executive procedures*: those which allow for the regular transaction of affairs within limits, and according to norms already established by a different authority.

4) *constitutive procedures*: those which create, extinguish or modify the exercise of power. Constitutive procedures comprise three subcategories:

i) *procedures of concession*: those which create new situations and new powers for the addressee(s).

ii)*procedures of limitation*: those which deprive physical or juridical persons of powers or faculties.

iii) *procedures of authorization*: those which consent to the exercise of powers already held by a physical or juridical person. They do not create powers, but remove limits to their exercise.[21]

---

21. Carucci, *Il Documento Contemporaneo*, p. 56.

These categories were identified by examining the documents issued by various medieval chanceries. On the basis of the apparent fact that documents had a different form depending on what they aimed to accomplish, it was assumed that the procedures generating the same documentary forms consisted of the same activities. As a logical consequence, to categorize documentary forms had to be equivalent to categorizing the procedures from which they derived, and therefore the activities generating them. So it was done. For example, it was found and established that the papal chancery issued privileges for conceding benefits (constitutive procedure of concession); *litterae gratiosae* for consenting to something (constitutive procedure of authorization); *litterae executoriae* for giving orders (constitutive procedure of limitation); *litterae concistoriales* for expressing collegial opinions (instrumental procedure); *litterae decretales* for formulating regulations (organizational procedure), etc. The results of this type of study were rationalized and generalized, and the operation brought about the categorization presented above.

Is this categorization applicable to modern procedures? How much has the world changed since medieval times? Can Montesquieu's trilogy of powers and its modern developments be compatible with the monolithic system viewed by diplomatists of medieval documents? This writer believes that the diplomatic categorization is valid with respect to modern procedures, although the world has undoubtedly become more complex, because human endeavours continue to present the same characteristics. In fact, individuals exist as human beings insofar as they belong to a group. Society gives itself a structure which regulates the coexistence of individuals, and establishes values and norms with which individuals wish to and must conform, and about which they share common ideas and opinions. It has been said that a collectivity founded on an organizational principle is a juridical system.[22] Within such a system, however much the governing principle changes over time and from place to place, human endeavours always present an *organizational*, an *instrumental* and an *executive* or

---

22. P. 61.

a *constitutive* nature. With respect to this categorization, one difference between the medieval and the modern worlds is that the four types of procedure can be found today at many levels rather than at one level only. This means that, from top to bottom, each category of procedure includes all the others, but, at any given level, confronted with a body of documents, we can say nevertheless what type of procedure is involved.

If we examine the three powers identified by Montesquieu, that is, legislative, executive or administrative, and judicial powers, we can see that, in democratic societies, they are primarily entrusted to separate bodies. This means that while each body exercises part of each power, its main competence is within the sphere of only one of them. Thus, for example, a parliament has primarily legislative competence; a government or a department has primarily administrative competence; a court has primarily judicial competence. As a consequence, we can say that legislative procedures are the constitutive procedures of a parliament; administrative procedures are the constitutive procedures of a department; and judicial procedures are the constitutive procedures of a court. However, each of these bodies functions by also carrying out the other three types of procedure.

It is the function of special diplomatics to focus on one specific records creator, study its procedures, categorize them according to the model proposed by general diplomatics, and proceed to further analysis. The latter, while aiming to group together similar procedures and to distinguish among different groups of procedures belonging to the same category, must proceed from the bottom up, that is, from the documents resulting from the procedures to the procedures themselves. For example, an exercise in special diplomatics intended to identify and reconstruct the procedures of Parliament, after having established that its constitutive procedures are those generating primary legislation, will be able to identify groups of procedures within the general constitutive procedure only on the basis of a categorization of laws. Laws can be categorized as "new laws," "major revisions," "major amendments," and "minor amendments," and further subdivided as "non-financial" and "financial," the former including "private bills" and "public bills," the latter comprising the three groups of laws related to "expenditures," "revenue" and "borrowing authority." Finally, special consideration should be given to "constitutional amendments." Clearly, the groups of procedures generating the categories of documents listed above have in common the fact of presenting a constitutive nature or purpose, and the fact of developing through the six phases described in the first part of this chapter. However, they differ within their constitutive purpose and within their phases of development. For example, the procedures originating "expenditure legislation" are procedures of authorization, while those creating "revenue legislation" are procedures of concession, which often arise out of the Budget Speech; "private bills" are introduced by private members, while "public bills" are usually introduced by the Cabinet; "new laws" often represent the culmination of a major government initiative, or the acceptance by the government of recommendations presented by a Task Force or Royal Commission of Inquiry, while "minor amendments" often result from the work of a technical committee.[23]

To understand the differences among these groups of procedures is essential for understanding the function and the intrinsic meaning of their documentary residue, and for evaluating it. However, such an understanding always begins with a direct examination of the documents embodying the procedures, and with an identification of their purpose. This direct examination and identification of purpose reveals another difference between the medieval and the modern worlds: whereas in the medieval context, each given documentary form was the result of one specific procedure and aimed at one specific purpose, in the modern context, procedures which are different, not as to their structure but as to their purpose, may create the same documentary forms; and, vice-versa,

---

23. Heather Heywood, Bob Krawczyk, Mary Ledwell, and Janice Simpson, "An Identification of Legislative Procedures," (paper prepared for the course ARST 601. Diplomatics; Master of Archival Studies, University of British Columbia, 1989). A study of special diplomatics on the organizational, instrumental, and executive procedures of the Parliament has still to be conducted, but it can be generally said that the procedures leading to the creation of internal rules meant to guide the conduct of parliamentary business have an organizational character, that those guiding the expression of opinions of committees and task forces have an instrumental character, and that those which constitute routines for the regular transaction of affairs have an executive character (i.e., procedure for the constitution of a technical committee). Studies of special diplomatics on the procedures of municipal

procedures having the same purpose may produce different documentary forms. However, this only reinforces the point made by diplomatics that documentary products must be "mapped" according to the functions and activities of their creators by reconstructing and examining the procedures of document-creation. This methodology of analysis, moreover, permits us to gain a knowledge of where and how information of a documentary nature can be shared among functions and juridical persons. For example, a series of retention and disposal schedules is a product of a constitutive procedure of authorization with respect to the archival institution which is competent for their approval, while it is a product of an executive procedure with respect to the document-creating agency which is competent for implementing them. The reconstruction of the typical procedure producing the schedules will allow us to identify the "workflows" which carry information both horizontally and vertically throughout an organization and between distinct organizations.

In conclusion, how can the diplomatic theory of document-creation provide a solution to the problem delineated by Gérard and Christiane Naud? How does it help us to understand whether a category of modern documents is equivalent to another category created in the past? This writer believes that diplomatic theory helps us by providing a method of analysis based on principles. The *principles* are:

1) Every procedure has the same ideal structure:

   i) The form of manifestation of the partial acts concurring with the main transaction is irrelevant to that structure, and so is the private or public nature of the juridical persons initiating and/or participating in the transaction.

2) All documents and the procedures generating them can be divided into categories on the basis of the purposes they were meant to accomplish:

---

departments and of provincial courts have been conducted in 1989 by two other groups of students enrolled in the Master of Archival Studies programme.

  i) The variation of the organizational principle on which a juridical system is founded is irrelevant to the general categorization, and so is the private or public nature of the juridical persons initiating and/or participating in the procedure.

The *method* is very familiar to archivists. Faced with a document or a group of documents (file, dossier, series), the archivist conducts his/her inquiry into its or their genesis from the point of view of the creator of the fonds to which the document(s) belong(s). When dealing with a single document, the archivist tries to identify, on the basis of its extrinsic and intrinsic elements of form and of its provenance,[24] its process of creation and the superior procedure in which it participated. When dealing with a group of documents, the archivist's inquiry will first be directed to the identification of those documents which participated in the same transaction, and then to the establishment of the procedural relationships existing among them, and of the analogous relationships between them and those documents in the same group which participated in other transactions. Afterwards, the archivist investigates how the group of documents under examination participated in superior procedures, and studies and categorizes them in absolute and contextual terms.

This method of analysis does not focus on subjects, but on actions of a very specific kind (i.e., initiative, inquiry, consultation, etc.); not on creating agents, but on creating procedures with defined purposes (i.e., organizational, instrumental, executive, etc.). The results of this focused analysis can then help to guide the efforts of appraisal, selection, arrangement and description, not only of the material analysed but also of all similar material. "What must be perceptible to those who read [archival descriptions], is the chain of the different stages of administrative action, the hierarchy of its

---

24. The extrinsic and intrinsic elements of form will be presented in the fifth chapter of this book. However, a basic introduction to those elements as clues to an understanding of procedures can be found in Janet Turner, "Experimenting with New Tools: Special Diplomatics and the Study of Authority in the United Church of Canada," *Archivaria* 30 (Summer 1990), pp. 91-103. The forms of transmission of documents are discussed on pages 49-53.

aspects and sub-aspects, the hierarchy of its purposes ... The vocabulary employed shall be coherent ... with regard to the actions."[25] This kind of study does not displace the traditional
archival inquiry into records creators, organizational structures
and subjects, but accompanies and complements it, just as the
diplomatic analysis of juridical systems supports the reconstruction of administrative histories, and the diplomatic examination of
physical and intellectual forms guides the study of content.

North American archivists instinctively have long recognized
the need to understand the routines governing creation of archives,
but only with the acquisition of electronic records, particularly
shared databases, has the central importance of procedure affected
their thinking. "An information system [writes John McDonald] is
a collection of records ... and processes, which are organized to
perform a specific set of functions in support of a defined set of
objectives."[26] The United Nations Advisory Committee for the
Coordination of Information Systems, moreover, points out, "Indeed, as we examine the electronic records landscape, it becomes
increasingly evident that the life cycle of the records (application)
system, and not the record, must be the new focus of attention. And,
on reflection, we can see that it was, or should have been, the focus
of attention in paper systems as well."[27] The terminology used by
electronic records specialists is very different from diplomatic terminology, but the message conveyed is clear: an understanding of
procedures is the key to the understanding of information systems.

It might be observed that, with regard to electronic systems, we
do not begin the analysis from the observation of the documents.
However, upon reflection, it is possible to see that we do. "Systems
developers are using tools and techniques that facilitate the design
of systems to manage the movement of (normally) structured

---

25. "Ce qui doit être perceptible à la lecture est l'enchaînement des diverses étapes
de l'action administrative, la hiérarchie de ses aspects et sous-aspects, la hiérarchie
de ses objects ... le vocabulaire employé devra être cohérent ... en ce que concerne les
actions." Naud, "L'analyse des archives ...," pp. 229 and 232.

26. John McDonald, "The Archival Management of a Geographic Information
System," *Archivaria* 13 (Winter 1981-82), p. 60.

27. *Management of Electronic Records: Issues and Guidelines* (New York: United
Nations, 1990), pp. 22-23.

information through pre-defined structured steps to achieve some pre-defined product (e.g., cheques, licences, etc.) in support of a programme activity."[28] Indeed, when we try to explore how the information system functions, we have in mind those pre-defined products; we know what the ultimate purpose of the system is. We can use the same approach suggested for paper systems. Close analysis of documentary products leads us to characterize the procedures by which they are created, on the basis of the ideal diplomatic procedure. We can then typify or generalize those procedures. At this point, instead of laboriously analysing every document to tease out from it an understanding of the procedures, we can begin by asking ourselves—knowing the kinds of possible procedures—what kind we face in any given instance.

It might also be observed that in many cases we already know the procedures from various sources such as annual reports, procedure manuals, policy files. But do we? These sources tell us how administrative action was supposed to be carried out, rather than how it actually was carried out; they tell us what the procedures ought to be, what management expected to happen, what the system was built for, and finally what the image was that the creating agency wished to reflect.[29] On the contrary, an analysis of the procedures which begins from their final products allows a verification of the discrepancies between rules and actuality and of the continuous mediation taking place between legal-administrative apparatus and society, and makes the reality attainable. This has always been the primary purpose of diplomatic analysis, and its value has not decreased. European archivists of the past generation used to teach their students to, "let the records tell their story," which may still be accepted as good advice by contemporary archivists. But, to understand that story, we may need a few more instruments. To provide them, the next chapter in this book will present the extrinsic and intrinsic elements of documentary forms.

---

28. John McDonald, letter to the writer, 25 September 1990.

29. About the importance of assessing the discrepany between image and reality by analysing the documentary product of activities, see Terry Cook, *The Archival Appraisal of Records Containing Personal Information: A RAMP Study with Guidelines*, (Paris, Unesco, 1990), in press.

# Chapter 5

# The Form of Documents and Their Criticism

Still glides the stream, and shall for ever glide
The form remains, the function never dies

Wordsworth, *The River Duddon*, 34,
"Afterthought"

The form of a document reveals and perpetuates the function it serves. On the basis of this observation, early diplomatists established a methodology for an analysis of documentary forms which permitted an understanding of administrative actions and the functions generating them. This methodology rested on the assumption that, notwithstanding differences in nature, provenance or date, all documents present forms similar enough to make it possible to conceive of one typical, ideal documentary form, the most regular and complete, for the purpose of examining all its elements.[1] Once the elements of this ideal form have been analysed and their specific

---

1. De Boüard writes that the analogous composition and the common traits of different documents are due to the fact that most documentary forms find their origin in the Roman *epistola*. Alain de Boüard, *Manuel de Diplomatique Française et Pontificale. Diplomatique Générale* (Paris, 1929), p. 255. Giry writes: "en dépit des différences du droit, des coûtumes et des usages, en dépit de nombreuses modifications dues aux circonstances particulières, aux influences locales, aux temps, ou même au caprice et à la fantaisie, il y a dans les chartes de toutes les époques et de tous les pays suffisamment de caractères communs pour qu'il soit possible d'en faire une étude méthodique." Arthur Giry, *Manuel de Diplomatique*. 1893. Reprint. (New York, n.d.), p. 481.

function identified, their variations and presence or absence in existing documentary forms will reveal the administrative function of the documents manifesting those forms.

Diplomatics defines *form* as the complex of the rules of representation used to convey a message, that is, as the characteristics of a document which can be separated from the determination of the particular subjects, persons or places which it concerns. Documentary form is both physical and intellectual. The term *physical form* refers to the external make-up of the document, while the term *intellectual form* refers to its internal articulation.[2] Therefore, the elements of the former are defined by diplomatists as external or *extrinsic*, while the elements of the latter are defined as internal or *intrinsic*.[3] From a conceptual point of view, it may be said that intrinsic elements of form are those which make a document complete, and extrinsic elements are those which make it perfect, that is, capable of accomplishing its purpose.[4]

This chapter will present and discuss the extrinsic and instrinsic elements of documentary form, and will show their relationship with administrative actions and functions.

## The Extrinsic Elements of Documentary Form

The extrinsic elements of documentary form are considered to be those which constitute the material make-up of the document and its external appearance. They can be examined without reading the document and are integrally present only in the original.[5] They are the medium, the script, the language, the special signs, the seals and the annotations. The study of these elements is properly the object of paleography, at least since the separation of this discipline

---

2. P. 41.

3. See: Giry, *Manuel*, p. 493; Alessandro Pratesi, *Elementi di diplomatica generale* (Bari, n.d.), p. 52; Paola Carucci, *Il Documento Contemporaneo. Diplomatica e Criteri di Edizione* (Roma, 1987), p. 98

4. Compare with the concept of originality as explained on page 49.

5. Pratesi, *Elementi di diplomatica*, p. 53.

from diplomatics formally took place in the nineteenth century.[6] However, diplomatics maintains its interest in them because the purpose of its analysis of those elements, namely the understanding of administrative processes and activities, is not directly pursued by paleography, which is more generally aimed at gaining an understanding of societal evolution, and of cultural, intellectual, ideological, economic and technical developments. Of course, diplomatics uses the intellectual instruments provided by paleography and other disciplines (e.g., sigillography) to analyse some of the extrinsic elements and their components, such as inks, illuminations, graphic characters and seals, but it only looks at specific aspects of them and for specific reasons. In fact, only certain parts of those extrinsic elements are especially relevant to diplomatics.

The first extrinsic element to consider is the *medium*, the material carrying the message. Traditionally, it has been essential for diplomatists to identify it (whether papyrus, parchment, paper, wooden tablet, etc.), to find out how it was prepared (e.g., the paste of the paper, the watermarks), and to note both its shape and size (or format) and the techniques used to prepare it for receiving the message (e.g., edging, ruling). This type of analysis was very important for medieval documents, because it made it possible to date them, establish their provenance, and test their authenticity. Later, much of its relevance was lost because offices were provided with their writing materials by manufacturing industries which served a great number of customers, and large bureaucracies adopted common materials. Today, with the increasing number of different types of physical media (e.g., magnetic tapes, optical discs), close attention to the medium chosen to carry a type of information can be very revealing of the ultimate purpose of that information—how it was meant to be used.

The other extrinsic element which used to have great significance for diplomatists, but progressively lost it, is the *script*. While it is the task of paleography to determine what type of script is proper to an era and an environment, it is the task of diplomatics to examine other characteristics of the script, such as the layout of

---

6. This point is specifically made by de Boüard *Diplomatique Française*, p. 224) and Giry (*Manuel*, p. 493).

the writing with respect to the physical form of the document, the presence of different hands or types of writing in the same document, the correspondence between paragraphs and conceptual sections of the text, type of punctuation, abbreviations, initialisms, ink, erasures, corrections, etc. With the invention of the printing press and, much later, of the typewriter, some of these characteristics became irrelevant to the purpose of diplomatic criticism. The need for careful examination of these characteristics is arising again, however, thanks to the advent of new technology. Computer software, for example, may be considered as part of the extrinsic element "script," because it determines the layout and articulation of the discourse, and can provide information about provenance, procedures, processes, uses, modes of transmission and, last but not least, authenticity. Also, the verification, editing and enhancement of the documentation associated with an electronic information system, carried out by the archivist who acquires the related electronic data files, is a modern form of diplomatic analysis of the script. Another example of such an analysis is the study leading to the definition of the Office Document Architecture (ODA) standard. The key feature of ODA is that it separates the "logical structure" of a document (i.e., paragraphs, sections and the relationship between them) from its "lay-out structure" (i.e., pagination, formatting), its "content" (in the technical jargon of the specialists, the way in which the message is represented: e.g., text and graphics), and its "profile" (which corresponds to the diplomatic "intrinsic elements"). The first three of the above four elements constitute those parts of the script with which diplomatics is concerned.

The *language* used in the document is an extrinsic element the importance of which is usually overlooked today, but which has been in the past the focus of attention by many diplomatists. Arthur Giry writes that because documents created in the course of administrative activity are destined to regulate interests,

the ideas expressed in them and the categories of facts to which they relate are necessarily limited in number, and recur very frequently in documents of the same type. Moreover, because it is important to discern easily the essential message within a document, ideas and facts are arranged in a given order which allows for ease of comprehension. Finally, because the expression and the organization [of

those ideas and facts] must be such that there will not be equivoca-
tions or misunderstandings, nor any need to refer again to the
subject, specific expressions and entire sentences are chosen and
made into formulas.[7]

Since the early Middle Ages, the art of composition and style was
the subject of regular instruction, which determined the development of a sort of documentary rhetoric, called *ars dictaminis* or
*dictamen*. Theorists established its rules, which were meant to direct
the composition, style and rhythm of every type of public document, private contract and business and family correspondence.
The various treatises which resulted used to be accompanied by
collections of models and examples, or of copies of actual documents, assembled for the purpose of showing the application of the
doctrine. These volumes, regularly used by public officers, notaries
and all those who needed to communicate in writing, were called
*formularia*.[8] Their production gradually diminished with the development of elementary education, but they are still created today for
the use of some professionals involved in the creation of types of
documents the language of which is highly standardized and controlled, such as lawyers.[9] With regard to electronic records, the

---

7. "... les idées qui y sont exprimées et les catégories de faits qui y sont relatées
sont nécessairement en nombre limité et se reproduisent assez fréquemment dans
les documents du même genre. De plus, comme il est important que l'on discerne
facilment dans un acte les dispositions essentielles, idées et faits y sont classés dans
un ordre combiné de manière à en rendre l'intelligence facile. Enfin, l'expression et
la disposition devant concourir à ce qu'il n'y ait ni équivoques, ni méprises, ni
malentendus, et à ce qu'on n'ait point à revenir sur les choses exprimées, il en est
résulté une recherche particulière d'expressions ou même de phrases entieres toutes
faites qui en constituent les formules." Giry, *Manuel*, p. 480.
    8. For an ample discussion of the *dictamen* and the *formularia*, see de Boüard,
*Diplomatique Française*, pp. 241-252, and Giry, *Manuel*, pp. 479-492.
    9. Some collections of copies of real documents have been assembled by diplo-
matists who, concerned with the absence or loss of *formularia* for some historical
periods, felt the need of having hand models to which they could compare the
various documents to be analysed and identified as to form and function. An
example is offered by Hubert Hall, *A Formula Book of English Official Historical
Documents*. 2 vols. 1908-9. Reprint. (New York, 1969).

codebook may be considered a modern *formularium* because of its instructional character.

The element of language is also studied, particularly by diplomatists of contemporary documents, from a social point of view. Different social groups use different forms of discourse and different vocabularies, and within each of them formal or informal styles are adopted, depending on the purpose and function of the documents created. There is no doubt about the existence of a curial, a journalistic, a political, a business, a scientific and a colloquial style. But it is important to underline that not just the style, but also the wording and composition of the documents created, for example, by a reporter, are radically different from those created by a lawyer, while those of a document created by a lawyer in carrying out his notarial function are different from those found in a lawyer's letter to a colleague.[10]

Among the extrinsic elements, diplomatists of medieval documents used to include the *special signs*, which should be regarded rather as intrinsic elements because of their function of identifying the persons involved in the documentation activity. The special signs can be divided into two categories: the signs of the writer and the subscribers, and the signs of the chancery or the records office. The first category includes the symbols used by notaries as personal marks in the medieval period, corresponding to the modern notarial stamp, and the crosses used by some subscribers in place of their name. The second category includes the *rota* and *bene valete* used by the papal chancery; the monogram of the sovereign's personal name used in imperial and royal chanceries; the initials *m.p.r.* for *manu proprio*; the double *s* for *s(ub)s(cripsi)*; and all the various office stamps.[11]

The most important extrinsic element of medieval documents, and the least common and relevant in contemporary documents, is the *seal*. Examining seals, diplomatists focus their attention on the

---

10. For a discussion of this issue see Carucci, *Il Documento Contemporaneo*, pp. 14-16.

11. Pratesi, *Elementi di diplomatica*, pp. 56-58. Giry considers the special signs to be an integral part of the validation of a document, and therefore discusses them in association with the subscriptions and signatures, that is, in the context of the "attestation," which is an intrinsic element of form. Giry, *Manuel*, p. 591.

material they are made of, their shape, size, typology (as it relates to the figure in the impression: heraldic type, equestrian, monumental, hagiographic, majestic, etc.), legend or inscription (the invocation, motto or title and name of the author, which runs clockwise around the central figure along the edge of the seal, starting from the top), and the method of affixing them (seals may be hanging or adherent). The analysis of these components is directed to ascertaining the degree of authority and solemnity of a document, its provenance and function, and its authenticity.[12]

The last extrinsic element to be considered, and the most relevant for contemporary documents, is the *annotations*. These can be grouped in three categories: 1) annotations included in a document after its compilation as part of the execution phase of an administrative procedure;[13] 2) annotations included in a complete and effective document in the course of carrying out the subsequent steps of the transaction in which the document participates; and 3) annotations added to a document by the records and/or archives service which is responsible for its identification as part of a group of documents (file, series) and for its maintenance and retrieval.

The main components of the first category of annotations include authentication and registration. *Authentication* may refer to one or more signatures, to an entire document, or to a copy of a document. It is the legal recognition that a signature is affixed by and belongs to the person whose name it expresses, that a document is what it purports to be, or that a copy conforms to the original.[14] *Registration* is the action of transcribing a document in a register, carried out by an office different from that issuing the document and specifically entrusted with that function. When registration takes place, the number assigned to the document in

---

12. For ample discussions of the seals from a diplomatic point of view, see Giry, *Manuel*, pp. 622-660, and de Boüard, *Diplomatique Française*, pp. 333-365.

13. The *execution phase* of an administrative procedure "is constituted by all the actions which give formal character to the transaction" (p. 116).

14. When the date of authentication is different from the date on which the document was compiled, and which appears among the intrinsic elements of documentary form, the former is considered to be the effective date of the document, for the legal purposes of avoiding fraud.

the register is included in the document with a formula attesting to that action. This formula and the registration number may be added to the document, not by the registration office, but by the notary or lawyer responsible for the compilation of the document, following proper authorization by the registration office.[15]

The second category of annotations comprises components such as question marks, initials, check marks and similar signs beside the text; indication of previous and/or following actions; dates of hearings or readings; notes of transmission to other offices; indication of future disposition; mention of the subject of the document; or locutions such as "Urgent," "Bring Forward," "Leave in abeyance," and so on.

The third category of annotations includes components such as the *registry number*, that is, the consecutive number assigned to incoming and outgoing mail in offices using the registry system; the *classification number*, which identifies a document and places it in relationship with those of the same transaction, file and series; *cross-references* to documents in other files and/or series; *date and office of receipt*; and *archival identifiers*, such as the consecutive page numbers given by an archives service, location codes, etc.

Annotations constitute the extrinsic element which most clearly reveals the formative process of a document, the way in which it participates in a transaction or procedure, and its custodial history.

To sum up, the extrinsic elements of documentary form as identified by diplomatics are the following:

| | |
|---|---|
| *Medium* | material |
| | format |
| | preparation for receiving the message |
| | layout, pagination, formatting |
| | type(s) of script |
| | different hands, typefaces or inks |
| | paragraphing |
| *Script:* | punctuation |
| | abbreviations and initialisms |
| | erasures and corrections |
| | computer software |

15. Registration is not a "formal" requirement for any document. For private documents, registration is only required for fiscal purposes, or for making the document public. Therefore, documents are "formally" complete and effective without registration.

|  |  |  |
|---|---|---|
| | formulae | |
| *Language:* | vocabulary | |
| | composition | |
| | style | |
| | signs of writers and subscribers | |
| *Special signs:* | signs of chanceries and record offices | |
| | material | |
| | shape and size | |
| *Seals:* | typology | |
| | legend or inscription | |
| | method of affixing | |
| | included in the | authentication |
| | execution phase | registration |
| | | signs beside text |
| | | previous or following |
| | | actions |
| | | dates of hearings or |
| | | readings |
| *Annotations:* | included in the | notes of transmission |
| | handling phase | disposition |
| | | subject |
| | | "Urgent" |
| | | "Bring forward" |
| | | registry number |
| | | classification number |
| | included in the | cross-references |
| | management phase | date and office of receipt |
| | | archival identifiers |

## The Intrinsic Elements of Documentary Form

The intrinsic elements of documentary form are considered to be the integral components of its intellectual articulation: the mode of presentation of the document's content, or the parts determining the tenor of the whole. The study of a great number of documents has shown that the elements which compose their intellectual form "are not simply juxtaposed, but tend to gather in groups, to be in some relationship of subordination one to the other, thereby forming sections each of which comprises several of them."[16] Therefore, it is possible to say that all documents "present an obvious typical

---

16. "les diverses parties qui composent un acte ne sont pas seulement juxta-posées, mais ... elles se groupent entre elles, ... elles se subordonnent en quelque sorte les unes aux autres, formant ainsi des divisions dont chacune comprend plusieurs des parties constitutives du document." Giry, *Manuel*, p. 527.

structure" and "an ideal analytical sub-structure."[17] This ideal
sub-structure comprises three sections, each of which has a specific
purpose. The first, termed *protocol*, contains the administrative
context of the action (i.e., indication of the persons involved, time
and place, and subject) and initial formulae; the second, termed *text*,
contains the action, including the considerations and circumstances which gave origin to it, and the conditions related to its
accomplishment; the third, termed *eschatocol*, contains the documentation context of the action (i.e., enunciation of the means of
validation, indication of the responsibilities for documentation of
the act) and the final formulae.[18] The three sections tend to be
physically distinct and recognizable, even in medieval and early
modern documents, which are not divided into paragraphs: usually, the three parts were identified by writing the first word of each,
and sometimes also the last, in a different script, style or dimension.

The intrinsic elements of form which usually appear at the
beginning of the document, that is, in its *protocol*, are numerous.[19]
Some of them are typical of medieval documents, others of contemporary ones; some are characteristic of documents issued by public
authorities, others of those issued by private juridical persons;
some belong in solemn documents, others in business documents;
finally, some are mutually exclusive while others tend to coexist.
They are described here in the order in which they appear when
they are all present.

In modern documents, at the very top we may have the *entitling*,
which today may correspond to the letterhead. It comprises the

---

17. "i documenti ... presentano una evidente struttura tipica ... una partizione
analitica ideale." Pratesi, *Elementi di diplomatica*, p. 62.

18. French and German diplomatists use the terms "initial protocol" and "final
protocol" for the first and third section of the document. The word protocol derives
from the Greek *protokollon*, which means "the first to be glued," and refers to the first
*plagula* or strip of the papyrus roll. Therefore, Italian diplomatists considered the
expression "initial protocol" to be a pleonasm, and the expression "final protocol"
to be a contradiction in terms, so they decided to call the first section simply
"protocol," and the third, by analogy, "eschatocol," from the Greek *eschatokollon*,
meaning "the last to be glued." Pratesi, *Elementi di diplomatica*, p. 63.

19. It may be interesting to note that the Italian register, in which the essential
data of incoming and outgoing documents are transcribed, is called "protocol." This
is probably a consequence of the fact that the data extracted from the documents for
registration are those contained in their protocol.

name, title, capacity and address of the physical or juridical person issuing the document, or of which the author of the document is an agent. Under the entitling or in its place we may give the *title* of the document (e.g., "Indenture," "Agreement," "Minutes," "This is the Last Will and Testament").

In contemporary documents, the entitling is usually followed by the *date*, indicating the place (*topical* date) and/or the time (*chronological* date) of the compilation of the document and/or of the action which the document concerns. In medieval and early modern documents the date is in the eschatocol.[20] In very solemn documents the date is present in both protocol and eschatocol.

The *invocation*, that is, the mention of God, in whose name each action had to be done, was present in both public and private documents in the medieval period. It can still be found in documents issued by religious bodies, but more and more rarely. When it appears, it takes a verbal form (starting with the words "in the name of") or a symbolic form (expressed by a cross, the Constantinian monogram for Christos, or the 'I' and 'C', for Jesus and Christus). The mention of God is in the eschatocol, when he is called to witness an act (e.g., an oath). It is possible to say that modern and contemporary documents contain an invocation whenever they present a claim that the act therein is done in the name of the people, the king, the republic, the law or other similar entities.

A typical element of the protocol used to be the *superscription*, that is, the mention of the name of the author of the document and/or the action. Today, the superscription tends to take the form of an entitling; sometimes, however, it coexists with the entitling. It still appears by itself in all contractual documents (the superscription includes the mention of the first party),[21] in declarative docu-

---

20. It is a fact that, over time, all elements connected to context have tended to move into the protocol, and the only elements left in the eschatol are the validation and some final clauses. With the evolution of technology, the validation has sometimes also moved into the protocol, and the subscription in the eschatocol appears more a formality than a real attestation; consider for example the telegram and the electronic mail. Independently of technology, some documentary forms tend to present an empty or almost empty eschatocol; consider for example the memorandum.

21. When analysing documents attesting to acts of reciprocal obligation, where each party is both author and addressee, diplomatists adopt the convention that the first party is the author and any other is the addressee. Hence, the name, title and address of the first party constitutes the superscription of every contractual docu-

ments (those beginning with the pronoun "I," followed by the name of the subscriber), and in holographic documents, such as wills (e.g., "This is the last will and testament [title] of John Smith of Vancouver" [superscription]).

Documents in epistolary form usually present in their protocol the name, title and address of the addressee of the document and/or the action. This element is termed *inscription*. It may be a *nominal* inscription or a *general* one. The former refers to one or more specific person(s), while the latter refers to a larger, indeterminate entity, such as the citizens, the people, the believers, the students, all those concerned, or "To all to whom these presents shall come." In contractual documents, given that the first party is considered to be the author, any other party is the addressee and the mention of his/her/their name(s) constitutes the inscription of the document. The inscription is regularly present in dispositive documents, often in supporting and narrative documents, but very rarely in probative documents, because usually the latter are not directed to the person to whom they are issued (e.g., certificates).

The inscription is generally followed by the *salutation*, a form of greeting which appears only in letters. In modern and contemporary documents the salutation is often in the eschatocol; sometimes it is in both the protocol and the eschatocol.

Today, the inscription may be followed by the *subject*, rather than by the salutation, that is, by a statement signifying what the document is about. The subject has been stated in some court records since the last century, but has generally been introduced into records of governmental bureaucracies and, by extension, into business records during this century.

Typical of medieval and early modern documents conferring titles or privileges is an element called *formula perpetuitatis*. It is a sentence declaring that the rights put into existence by the document are not circumscribed by time: *in perpetuum* (forever), *ad perpetuam rei memoriam* (in continuing memory), or *pp.* (abbreviation of *perpetuum*).

---

ment (p. 104).

Another medieval formula is the *appreciation*, that is, a short prayer for the realization of the content of the document: *feliciter* (happily), or *amen* (so be it). It appears in the protocol in private documents, and in the eschatocol in public documents following the date. A modern form of appreciation may be considered to be the expression which often concludes contemporary documents, and which is introduced by the words "looking forward to," "I appreciate," "I hope," etc.

The *text* is the central part of the document, where we find the manifestation of the will of the author, the evidence of the act, or the memory of it. From an historical, legal and administrative point of view this is usually the most important part of the document, because it represents its substance, the reason for its existence. However, to the diplomatist, the text does not offer more material for the criticism of the document than the other two sections.

The text often begins with a *preamble*, which expresses the ideal motivation of the action. It does not give the concrete and immediate reason for which the document was created, or the action accomplished, but the ethical or juridical principle. It consists of general considerations, which are not directly linked to the subject of the document, but express the ideas which inspired its author. The preamble has the purpose of engaging the addressee's interest and ornating the discourse, and is therefore composed of moral or pious expressions, sentences expressing political conceptions, administrative policies, legal principles, feelings of friendship, cooperation, interest, security, and so on. The preamble has never been an essential part of the text, thus its presence indicates solemnity or formalism. In modern legal documents, the preamble contains a citation of the laws, regulations, decrees, or opinions on which the act rests. Today, just as in the past, it is possible to notice that some types of documentary form have their own specific, and often stereotyped preamble. "When this part of the text is not copied from ancient formularia or previous acts, one recognizes in it, better than in any other part, the mark of an epoch, the characteristics typical of certain categories of acts or of certain chanceries, and also the imprint of the personality of its author. The ideas themselves which are expressed in the preamble can serve in some measure as elements of criticism."[22] (For example, in royal letters patent of appointment, the preamble reads: "Whereas We have taken into Our Royal Consideration the Loyalty, Integrity and Ability of Our Trusty and Well-beloved ...").

In some official dispositive documents the preamble is followed by the *notification*, that is, by the publication of the purport of the document. Its purpose is to express that the act consigned to the document is communicated to all those who have interest in it and, as well, that all persons concerned must be aware of the dispositive content of the document. The notification consists of a formula, such as *"notum sit,"* "be it known," "know you," and sometimes commences the text and is followed by, or exists without the preamble.

The substance of the text is usually introduced by the *exposition*, that is, the narration of the concrete and immediate circumstances generating the act and/or the document. In documents resulting from procedures, whether public or private, the exposition may include the memory of the various procedural phases, or be entirely constituted by the mention of one or more of them. Thus, in documents conceding something, there is a mention of the request, of the reasons for the request and for its acceptance, and of the consensus and advice of the interested parties; in documents relating to contentious acts, there is the history of the case and its development; in warrants, we find a narration of facts, circumstances, reasons determining the decision, and so on.[23] Sometimes the exposition includes names of individuals who have participated in the decision-making process, such as intermediaries, advisers, friends or relatives. It happens that many documents, both public and private, originate from analogous situations. In these cases, the narration becomes a stereotyped formula which, in legal documents, especially in those of a contractual nature, is prescribed

---

22. "Lorsque cette partie du texte n'a pas été recopiée sur d'anciens formulaires ou sur des actes antérieurs, ou y reconnaît, mieux que dans aucune autre, la marque d'une époque, des caractères particuliers à certaines catégories d'actes ou à certaines chancelleries, et même l'empreinte de la personnalité de son auteur." Giry, *Manuel*, p. 543.

23. For a discussion of the phases of a procedure, see pp. 115-116.

by law. In contemporary documents, such a formula is usually preprinted formally, and begins with "whereas."

The core of the text is the *disposition*, that is, the expression of the will or judgement of the author. Here, the fact or act is expressly enunciated, usually by means of a verb able to communicate the nature of the action and the function of the document, such as "authorize," "promulgate," "decree," "certify," "agree," "request," etc. The verb may be preceded by a word or locution which puts the disposition in direct relationship to the previous exposition or preamble, such as "therefore," "hereby," etc. There are specific formulae routinely used for certain types of transaction, but generally the disposition varies from one document to another because there are no two acts which are quite the same.

In many documents the text ends with the disposition, that is, as soon as the substance of the action is expressed. The text of most documents, however, contains after or within the disposition several formulae, the object of which is to ensure the execution of the act, to avoid its violation, to guarantee its validity, to preserve the rights of third parties, to attest the execution of the required formalities, and to indicate the means employed to give the document probative value. These formulae constitute the *final clauses* which can be divided into groups as follows:

*Clauses of injunction*: those expressing the obligation of all those concerned to conform to the will of the authority.

*Clauses of prohibition*: those expressing the prohibition to violate the enactment or oppose it.

*Clauses of derogation*: those expressing the obligation to respect the enactment, notwithstanding other orders or decisions contrary to it, opposition, appeals or previous dispositions.

*Clauses of exception*: those expressing situations, conditions or persons which would constitute an exception to the enactment.

*Clauses of obligation*: those expressing the obligation of the parties to respect the act, for themselves and for their successors or descendants.

*Clauses of renunciation*: those expressing consent to give up a right or a claim.

*Clauses of warning*: those expressing a threat of punishment should the enactment be violated. They comprise two categories: 1) *spiritual sanctions*, comprising threats of malediction or anathema; 2) *penal sanctions*, comprising the mention of specific penal consequences.

*Promissory clauses*: those expressing the promise of a prize, usually of a spiritual nature, for those who respect the enactment.[24]

*Clauses of corroboration*: those enunciating the means used to validate the document and guarantee its authenticity. The wording changes according to time and place, but these clauses are usually formulaic and fixed. Examples are "I have hereunto set my Hand and Seal of Office," "Signed and Sealed," "Witness our Trustworthy and Beloved ...," etc.[25]

More and more often, particularly in solemn, official and legal documents, the clause of corroboration begins the *eschatocol*, immediately followed by the topical and chronological date, or a reference to the date expressed in the protocol (e.g., "In Testimony whereof I have hereto set my hand and seal at Johnstown aforesaid this fourth day of July in the year of our Lord one thousand eight hundred and eight," and "In Witness Whereof, the said parties have hereunto set their hands and seals, the day and year first above written"). In non-official documents, and in documents of private

---

24. The clauses of warning and the promissory clauses are called by some diplomatists, respectively, negative sanctions and positive sanctions.

25. For a discussion in depth of the final clauses, see Giry, *Manuel*, pp. 553-572, and de Boüard, *Diplomatique Française*, pp. 277-292.

origin, the eschatocol may begin with a sentence of appreciation, followed by the salutation, and by the *complimentary clause*, which consists of a brief formula expressing respect, such as "sincerely yours," "yours truly," and similarly. Whatever the case, the substance and core of the eschatocol is the *attestation*, that is, the subscription of those who took part in the issuing of the document (author, writer, countersigner) and of witnesses to the enactment or the subscription. Usually, the subscription takes the form of a signature, but this is not always so; for example, telegrams and electronic mail messages present subscriptions which are not signatures. The attestation is the means generally used to validate a document, but is not present in every type of document. For example, account books, journals and invoices do not need a subscription to be valid because their process of creation validates them. Other documents present their validation in the protocol. This is typical of electronic records, but examples can also be found in traditional records: registries may be validated on the front page, memoranda may be signed or initialled on the side of the superscription, and documents issued by the English monarchs show the *signum manus* in the top left corner. A discussion of the various types of attestation, their meaning and their function, is not among the purposes of this chapter and would warrant the space of an entire chapter.[26]

When the attestations are signatures, they are usually accompanied by the *qualification of signature*, that is, by the mention of the title and capacity of the signer. The qualification of signature may be followed by the *secretarial notes* (initials of the typist, mention of enclosures, indication that the document is copied to other persons, etc.), but usually it constitutes the last intrinsic element of documentary form.

To sum up, the intrinsic elements of documentary form are:

> entitling
> title
> date

---

26. For a discussion of the various signs of validation of a document, see Giry, *Manuel*, pp. 591-621, and de Boüard, *Diplomatique Française*, pp. 321-333. For the identification of the persons signing a document, see pp. 82-95 of this volume.

|              |                        |
|--------------|------------------------|
| *Protocol:*  | invocation             |
|              | superscription         |
|              | inscription            |
|              | salutation             |
|              | subject                |
|              | *formula perpetuitatis*|
|              | appreciation           |
|              |                        |
|              | preamble               |
|              | notification           |
| *Text:*      | exposition             |
|              | disposition            |
|              | final clauses          |
|              |                        |
|              | corroboration          |
|              | [date]                 |
|              | [appreciation]         |
| *Eschatocol:*| [salutation]           |
|              | complimentary clause   |
|              | attestation            |
|              | qualification of signature |
|              | secretarial notes      |

The intrinsic elements listed above do not appear all at the same time in the same documentary form, and some of them are mutually exclusive. According to Hubert Hall, a typical English official document of the medieval period is composed as follows:

|              |                        |
|--------------|------------------------|
| *Protocol:*  | invocation             |
|              | superscription         |
|              | preamble               |
| *Text:*      | exposition             |
|              | disposition            |
|              | final clause of warning|
|              |                        |
|              | date                   |
| *Eschatocol:*| attestation[27]        |

However, it is the specific combination of those elements which determines the aspect of documentary forms, and allows us to distinguish one form from another at a glance.

---

27. Hubert Hall, *Studies in English Official Historical Documents.* 1908. Reprint. (New York, 1969), pp. 190-192.

## The Structure of Diplomatic Criticism

The extrinsic and intrinsic elements of documentary form were identified by diplomatists through examining a great number of documents issued in different times and jurisdictions by different types of records creators for different purposes. The immediate aim of such identification was to put into direct correspondence the single components of documentary form with specific components of the administrative transaction, and the various combinations of those components with given types of transaction. Its ultimate purpose was to achieve the ability to see the function of documents through their form, to learn about functions as they were accomplished by each records creator, and thus to gain the knowledge necessary to verify the authenticity of documents which purport to have been created by a given juridical person while carrying out a specific function.

Diplomatic criticism therefore proceeds from the form of the document to the act initiated or referred to by the document. This analysis aims at understanding the juridical, administrative and procedural context in which the documents under examination were created.

The structure of diplomatic analysis is quite rigid and reflects a systematic progression from the specific to the general. This is the only direction which can possibly be taken when the context of the document under examination is unknown. Therefore, diplomatic criticism proceeds as follows:

| | |
|---|---|
| *Extrinsic elements:* | medium |
| | script |
| | language |
| | special signs |
| | seals |
| | annotations |
| *Intrinsi elements:* | protocol |
| | subsections |
| | text |
| | subsections |
| | eschatocol |
| | subsections |
| *Persons:* | author of the act |

author of the document
addressee of the act
addressee of the document
writer

countersigner(s)
witness(es)

*Qualification*            titles and capacity of the persons involved
*of signatures:*

*Type of act:*             simple, contractual, collective, multiple, continuative, complex
                           or procedural

*Name of act:*             e.g., sale, authorization, request

*Relationship*             specification of the phase of the general procedure to which
*between documen*           the documents relates and, if the document results from an
*and procedure:*           "act on procedure," the phase of the specific procedure

*Type of document:*        name (e.g., letter, indenture)
                           nature (public or private)
                           function (dispositive, probative, etc.)
                           status (original, draft, or copy)

*Diplomatic:*              context (year, month, day, place)
*description:*             action (persons, act)
                           document (form name, nature, function, status,
                           medium, quantity)

*Conclusive*               any comment which would refer to the document as a
*comments:*                whole rather than to a specific element of documentary
                           form or component of diplomatic analysis[28]

As a demonstration of how diplomatic criticism of documentary forms is conducted, two documents are now analysed according to the pattern delineated above. This analysis is not complete because the extrinsic elements of documentary form can only be criticized on the basis of the original document.[29] However, the extrinsic

---

28. In this rigid model, comments referring to single elements of the documentary form under examination or to single components of the diplomatic analysis are offered in footnotes. These are identified by letters if the comments they contain are of a diplomatic nature, and by numbers if the comments are of a historical-juridical nature.

29. Even if the originals of the documents criticized below were available to the author, they would not be to the readers, so it appears to be a useless exercise to comment on something which cannot be seen.

# PISCATAWAY INDIAN NATION

☩ : Andrew WHITE. *open in bryn... norman nepli under d'Rom*

June 18, 1980

Archivio Cenprale Dello
Stato Piazzale Archivi (EUR)
Rome, Italy

To whom it may concern:

The Piscataway Indian Nation is the native people to the State of
Maryland. We are a poor people trying to maintain our culture,
heritage and identity.

According to information we have, a Jesuit Missionary, Father Andrew White, composed a catechism in the native dialect of the Piscataway Indians. He also compiled a grammar and dictionary in the
Indian language. The catechism is reported to have been printed
on one of the first printing presses in the colonies and was discovered years later in the Archives in Rome.

To go back a little, Father White along with several other Jesuits
were one of the first to come to what is now known as the State of
Maryland. Before long he had baptized my people into the
Catholic faith, of which our people are still devout.

What we would like to have is, if possible, copies of the Catechism,
grammar and dictionary.

We understand that there is more than one Archives in Rome. If you
are unable to help us in this matter, we would appreciate a listing
of other Archives in Rome. We appreciate your time and effort in
this matter and look forward to hearing from you.

May Mother Earth Endure Her Suffering.

Chief Billy Redwing Tayac
Piscataway Indian Nation

Figure 3

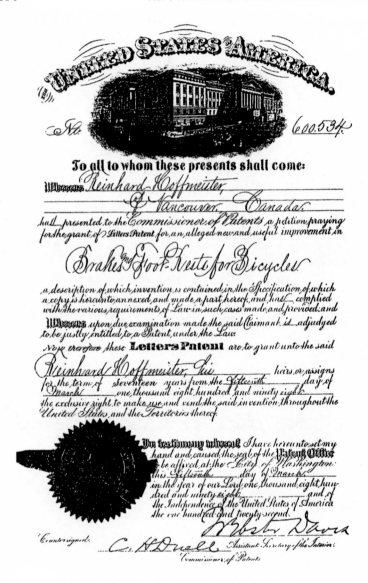

Figure 4

elements which are essential to the understanding of the actions in which those documents participated are mentioned in the context of the analysis of those actions. No indication is given of the provenance of the two documents, in order to show more clearly the perspective of the diplomatists who devised this method of analysis.

## Diplomatic criticism of the document in Figure 3

### Intrinsic elements

| | |
|---|---|
| *Protocol:* | "Piscataway ... concern" |
| | entitling: "Piscataway ... nation" and Insignia |
| | chronological Date: "June ... 1980" |
| | inscription: "Archivio ... concern" |
| *Text:* | "The Piscataway ... hearing from you" |
| | preamble: "The Piscataway ... identity" |
| | exposition: "According ... devout" |
| | disposition: "What ... dictionary" and "We ... other archives in Rome" |
| | appreciation: "We appreciate ... from you" |
| *Eschatocol:* | "May ... Nation" |
| | appreciation: "May ... Suffering"[30] |
| | attestation: "Billy Redwing Tayac" |
| | qualification of signature: "Chief ... Nation" |
| *Persons:* | author of the act: Piscataway Indian Nation |
| | author of the document: Piscataway Indian Nation |
| | addressee of the act: Archivio Centrale dello Stato (Rome) |
| | addressee of the document: Archivio Centrale dello Stato |
| | writer: Chief Billy Redwing Tayac[31] |
| *Qualification of signature:* | Chief of the Piscataway Indian Nation |
| *Type of act:* | simple act |

---

30. The appreciation is defined as a wish or prayer for the realization of the intention of the document. This document presents two appreciations, one of which is expressed at the end of the text in modern style, and the other at the beginning of the eschatocol in the traditional form of invocation. In formal diplomatic criticism, this comment, being of diplomatic nature, would be introduced by a letter. In the present context, this is avoided, so as not to create confusion.

31. The reasoning behind the identification of the persons is illustrated on pp. 89-91.

| | |
|---|---|
| *Name of act:* | request for information |
| *Relationship between document and procedure* | document participating in the initiative phase of a compound act on procedure[32] |
| *Type of document:* | letter; public; dispositive; copy[33] |
| *Diplomatic description:* | 1980, June 18. [Maryland, U.S.A.]. The Piscataway Indian Nation asks the Archivio Centrale del Stato in Rome for copies of a catechism, grammar and dictionary in its native dialect. |

1 letter, public, dispositive, copy
(A.D. 1980, June 26. Rome, Italy)[34]

## Diplomatic criticism of the document in Figure 4

### Intrinsic elements

| | |
|---|---|
| *Protocol:* | "The United States of America ... come" entitling: "The United States of America" title: "No ... 600.534" (patent number) |

---

32. For the definitions of *simple act* and *compound act on procedure* see pp. 75-76. For the definition of *initiative phase* see p. 115. Both the simple act of requesting information and the procedure of providing reference service are juridical acts, because their consequences are taken into consideration by the juridical system in which they take place. If the addressee of the request did not act on it, he would have incurred "neglect of an official duty."

33. The annotations in the document show that it was received by the addressee, registered, classified, and then passed to the competent person for action. The subject of the document is handwritten in Italian, and the author of the document is pointed to by an arrow, probably to emphasize the two elements essential to the accomplishment of the service. This document is a letter because the tenor of the discourse is modelled on the classic *epistola*, presents subjective wording (the author is in the first person), and its addressee is identified. It is public because it participates in a public procedure (the Archivio Centrale dello Stato is a public institution where reference service is mandated by an act of law). It may also be argued that its author is a public entity within the Indian juridical system (see pp. 102-105). This document is dispositive because it puts the act of request into existence (of course, it is dispositive only in its original status).

34. When date of receipt is known, it is usually added to the "document area" of the diplomatic description, preceded by the initials a.d. (archival date).

inscription: "To all ... come"

| | |
|---|---|
| *Text:* | "Whereas ... Thereof"<br>exposition: "Whereas ... Law"<br>disposition: "Now therefore ... thereof" |
| *Eschatocol:* | "In testimony ... Commissioner of Patents"<br>corroboration: "In testimony ... affixed"<br>topical date: "at the City of Washington"<br>chronological date: "this fifteenth ... second"<br>attestations: 2 signatures<br>qualifications of signature: "Assistant Secretary of the Interior" "Commissioner of Patents" |
| *Persons:* | author of the act: The United States of America<br>author of the document: The United States of America<br>addressee of the act: Reinhard Hoffmeister of Vancouver<br>addressee of the document: "To all to whom these presents shall come"<br>writer: the Assistant Secretary of the Interior, Davis<br>countersigner: the Commissioner of Patents, Duell[35] |
| *Qualification of signatures:* | Assistant Secretary of the Interior, Commissioner of Patents, Duell |
| *Type of act:* | simple act |
| *Name of act:* | granting of a patent for an invention |
| *Type of document:* | letters patent, public, dispositive, copy[36] |
| *Diplomatic description:* | 1898, March 15. Washington, D.C.<br>The United States of America grants Reinhard Hoffmeister of Vancouver, Canada, a patent for an invention.<br><br>1 letters patent, public, dispositive, copy |

35. This countersignature has the function of attesting the regularity of the procedure of formation and of the forms of the document, while the signature of the writer attests to the fact that the action in the document conforms to the will of the authority.

36. A *letters patent* is an instrument proceeding from a sovereign authority, and conveying a right, authority or grant to an individual. It is in the form of a letter delivered open, not closed up from inspection like the *letters close*. In fact, the content of a letters patent is meant to be known to all those concerned.

*Relationship*            document concluding the execution phase of a
*between*                 compound act on procedure[37]
*document and procedure:*

The diplomatic criticism conducted above may seem a sterile exercise of identification and "labelling."[38] However, the exercise itself is the key to an understanding of the action in which the document participates, and of the document itself. The names on the labels are indicators which direct attention to the entities which are relevant to the continuous process of extrapolation by the archivist. The effort of including the elements of real documents into the framework of diplomatic analysis is a necessary prelude to discovery and knowledge. One might object that archivists do not describe single items. That is not necessarily the case. When they do not, it is because they are already familiar with them—culturally familiar—and the process of extrapolation takes place spontaneously.

Diplomatics is a mind-set, an approach, a perspective, a systematic way of thinking about archival documents. How to make the best use of its concepts and methodology in archival descriptive work will be the subject of the next chapter, the sixth and last in this book.

---

37. For the definition of *execution phase* see p. 116.

38. This term is used by Janet Turner in the comments following her diplomatic analysis of three documents of the United Church of Canada ("Experimenting with New Tools: Special Diplomatics and the Study of Authority in the United Church of Canada," *Archivaria* 30 (Summer 1990), p. 99). Turner's article is useful reading for all those interested in the use of diplomatic criticism.

# Chapter 6

# The Uses of Diplomatics

"After all is said and done, it is the record which is our special area of knowledge."

Barbara L. Craig[1]

Over the centuries, the focus of diplomatics has remained the archival document, the record embodying action. Diplomatists have dissected it in its constituent parts and observed it as a whole; they have linked it to act, procedure, person, function, system and analysed its relationships with those entities; they have studied its causes and effects, its reality and the idea behind it, and its individuality, and context.

In the nineteenth century, European archivists, recognizing the archival document as the focus of their scholarly activity, and the knowledge of it as the intellectual foundation of their institutional and professional functions, included diplomatics among the sciences constituting the core curriculum of the schools created to educate members of their profession. Today, diplomatics remains a fundamental subject in all European archival schools; its relevance and its formative function in the education of archivists are not questioned; its usefulness for the identification and control of archival documents of the past centuries is a proven fact; yet doubts

---

1. Barbara L. Craig, "The Acts of the Appraisers. The Plan, the Context and the Record." Commentary on the paper by Hans Booms, "Überlieferungsbildung: Archives-Keeping as a Social and Political Activity," delivered at the Annual Conference of the Association of Canadian Archivists (Banff, 23 May 1991), p.11.

arise in the minds of those archivists who work only with modern documents about the direct applicability of its methods and the use of its concepts.

After World War II, the dramatic increase in document production obliged archivists to shift the focus of their attention from the document to larger and larger groups of documents. With modern records, physical arrangement of items has given way to intellectual arrangement of files and series;[2] calendars and analytical inventories have been abandoned in favour of inventories at the series level or even summary inventories; appraisal has gone from the weeding of duplicates and ephemera (destruction of the redundant and useless), to identification of the series to be preserved (selection of the significant and useful); and, most importantly, the analysis of the archivist is gradually moving from the immediate documentary context of the material under examination to its broad functional context and, further, to its socio-cultural context; that is, from the reality of the records to the "image" of records creators.[3] Moreover, electronic information systems are producing a records reality apparently so different from the one archivists are used to seeing that it is difficult for them to believe that there is a record reality at all: virtual documents, dynamic documents, compound documents, smart documents, hyperdocuments, documentary views—even the names reflect a sense of uncertainty, instability, and confusion.[4]

What is the role of diplomatics in all this change? Does it indeed have a role? The answer can hardly come from Europe, because

2. Of course, intellectual arrangement was not a new concept, a surrogate of physical arrangement, because it had always preceded it, but the idea that "walking along the shelves must be like walking along history"—as Francesco Bonaini, the Archivist of the Grand Duchy of Tuscany, put it—was completely abandoned.

3. A discussion of the approach which focuses on the "image" of society reflected by records-creating bodies can be found in Terry Cook, *The Archival Appraisal of Records Containing Personal Information: A RAMP Study with Guidelines* (Paris, 1990), p.41 ff.

4. More and more often archivists are expressing their frustration by wishing there was a firm definition of what an archival document or a record is. In fact, the most recent reports issued by archival committees start with an attempt to formulate such a definition. One example is *Management of Electronic Records: Issues and Guidelines* (New York, 1990), p.20.

European archivists are all educated in diplomatics: its principles, concepts and methods are an integral part of their mind-set and outlook. If they are not fully aware of how a knowledge of diplomatics contributes to their work with modern records, they certainly would not know whether its absence would make such work more difficult and, if so, in what manner. Rather, the answer can come from North American archivists, for whom diplomatics is a recent discovery in two ways: one direct, the other indirect. On the one hand, they can try to apply diplomatic concepts in the course of their work; on the other hand, they can observe more closely than they have done so far the work of European archivists who are dealing with the same issues and problems with which North American archivists are confronted, but are using different approaches which sometimes have a clear diplomatic matrix.[5] The first way seems to be the more difficult at the moment, because this book is the first exposition of general diplomatics in the English language, and its content appears to be very abstract, if not exotic. This final chapter will try to overcome the difficulty by explaining how the concepts illustrated in the previous five chapters can be useful to North American archivists, that is, how they can inspire and permeate their work. The Appendix contains a list of the

---

5. A consequence of the fact that European archivists are not fully aware of the contribution of diplomatics to their work is that, in the last twenty years, with a few notable exceptions, they have not articulated this issue in their writings. Moreover, some European archivists have expressed perspectives which may appear to over-rule diplomatic methods in the accomplishment of the major archival functions, simply because they tend to emphasize what has still to be achieved rather than what is already an integral part of their cultural background (see, for example, the writings of Hans Booms, Siegfried Buttner and Michael Cook). For this reason, I suggest that North American archivists go beyond reading what European archivists write by observing their work and, when possible, working with them. This kind of observation, comparison and cooperation has already began, if only on a very small scale. Expressions of it are an article by David Bearman and Peter Sigmond, "Exploration of Form of Material Authority Files by Dutch Archivists," *The American Archivist* 50, 2 (Spring 1987), pp. 249-253; the invitation to Peter Sigmond by the Association of Canadian Archivist to its Annual Conference in Banff, Alberta (21-25 May 1991), where he presented a paper on form, function and archival value; and the meeting of specialists on "Electronic Records and Archival Theory," held in Macerata, Italy (13-17 May 1991), the final report of which, written by Charles Dollar, will be published by the University of Macerata, in Italian and English, in 1992.

diplomatic definitions presented in this book, which guides the
reader to the page on which they are offered.

## Diplomatics as a Formative Discipline

It has been repeatedly said that for the archivist diplomatics is a
formative discipline.[6] Its function is the same as anatomy for the
medical doctor, physics for the engineer, and grammar for the
linguist or any literate person. The analogy between diplomatics
and grammar is particularly evident, not only as relates to the
structure and function of the discipline, but also with regard to its
evolution. In the Middle Ages, grammar was one of the seven
liberal arts taught in the monastic and cathedral schools, and later
in the universities.[7] Over time, its centrality in the education of the
man of letters was usurped by less analytic and more holistic
disciplines; thus, grammar gradually disappeared from the general
curricula of study, to be almost exclusively relegated to the realm
of the linguist. The times of the "grammar, which can govern even
Kings" seem to be gone forever.[8] Is this a good thing? Few people
would say so. Grammar allows for easy and accurate communication by making explicit the set of principles by which a language
functions. By learning those principles, the persons who share the
same language share a single standardized system of communication. Such a system includes traditional grammar, which defines

---

6. For example, see Cencetti's definition of diplomatics, which opens this book.
Addressing the issue of archival education, James O'Toole writes "we have been less
interested in teaching students to think like archivists than we have in getting them
to act like archivists," and "our concern has been with what an archivist can be
trained to *do*, rather than with what an archivist should be educated to *know*"
["Curriculum Development in Archival Education: A Proposal," *The American Archivist* 53 (Summer 1990), p.463]. Diplomatics is considered to be a formative
discipline precisely because it instils into archival students a way of thinking and a
specific knowledge which do not belong to any other profession and are therefore
characteristic of the archival profession.

7. The seven liberal arts were grouped in the trivium (grammar, rhetoric and
logic) and the quadrivium (arithmetic, astronomy, geometry and music).

8. "La grammaire qui sait régenter jusqu'aux rois," Molière, *Les Femmes Savantes*,
II. 6.

parts of speech by their meaning and function; structural grammar, which defines them primarily by their order in a sentence; and transformational grammar, which moves the emphasis from analysis of the parts of speech to the way people produce all the possible sentences of the language.

At this point, the parallel between grammar and diplomatics should be clear. The first important contribution of diplomatics to archival work is its definitional component, which identifies the meaning and function of the constituent parts of the document, and names them in a consistent and significant way. This is not a minor thing, not only from a communication point of view but also from a standardization point of view. Modern archivists use terms such as "medium," "form," "logical relations," "physical relations," "logical structure," "layout structure," "document profile," in a very inconsistent way, and keep creating arbitrary terms every time they encounter an entity which appears slightly different from those with which they are familiar. Failure to recognize the substance of things leads to the false impression that the reality is changing fundamentally, and this generates panic in those who have to deal with it. The precision of diplomatic terminology gives communication between archivists and among the information professions a clarity which is lacking in much of the terminology currently in use. For example, the term "hypermedia documents" is used to refer to documents which differ as to information configuration (i.e., the main attribute of the script: text, graphic or image, what David Bearman calls "sensory information modality"), format or layout, and intellectual forms, but which, at the creation stage, are all stored in the same medium and linked together as elements in a Hypertext system. As another example, the term "textual documents" is usually contrasted with the term "electronic documents," when "textual" connotes an information configuration and "electronic" a method or agency of creation, preservation and transmission; it is a fact that a textual document electronically stored in a magnetic tape (the carrier or medium) remains a textual document, while an electronic document is a document electronically created, maintained, or transmitted independently of the configuration of the information which it contains.[9]

These examples show that there is a very real risk that the method of transmission of documents will become the paramount element in archival discourse, to the point that a facsimile is already considered by many to be an electronic document on the grounds that it is electronically transmitted. Using such standards of terminology, we should have to call all documents delivered by regular mail "postal documents" and those delivered by hand "courier documents"!

Diplomatics has always maintained a distinction between the "method of transmission" of a document and its "form of transmission." The latter refers to the information carrier, or medium, on which the document is received by the addressee, and therefore, with respect to electronic transmission, to the physical final product of that operation. Thus, diplomatically, a facsimile received on paper is a manuscript and must be treated as such, while one received on a computer screen is electronic and must be treated as a computer document.[10]

---

9. For a discussion of the terminological confusion which surrounds the entire area of records electronically generated, see Catherine Bailey, "Archival Theory and Electronic Records," *Archivaria* 29 (Winter 1989-90), pp.181-2; and "Archival Theory and Machine Readable Records: Some Problems and Issues" (M.A.S. Thesis, University of British Columbia, 1988), pp. 5-25. To say that diplomatics helps to clarify the terminology currently used is not to suggest that archivists should substitute the commonly accepted terms with diplomatic terms; it rather means that diplomatics explains the nature of the entities we refer to by various, often inconsistent terms. Thus, one can continue to call manuscripts "textual documents" so long as one is aware that this is a conventional term, not a substantial one; or, one can list maps among other media so long as one knows that the term refers to an information configuration in the same category as a chart or a plan, which may be stored in any medium.

10. This is also the point of view of Fred V. Diers, who believes that the method of transmission should not be taken into account when defining a medium: information storage is the first definable element of the "matrix." Fred V. Diers, "The Information Media Matrix: A Strategic Planning Tool," *Records Management Quarterly* (July 1989), pp. 17-23. David Bearman, on the contrary, believes that the method of transmission does change diplomatic forms substantially in the case of electronic mail, because "the manifestation of electronic mail is different to the sender, the system and the recipient in a way that has no analog in paper based communications" (Letter to the author, 20 October 1991). I believe that such a difference has been

Diplomatics also distinguishes between the "method of transmission" and the "status of transmission" of a document, the latter
being its degree of perfection.[11] For example, with regard to its
status of transmission, a facsimile on its own is an "imitative copy."
It may be certified as an authentic copy, that is, a copy with the
validity and effect of an original, by its method of transmission, if
the technology used can guarantee its trustworthiness. However,
technological devices can only ensure that the facsimile conforms
to the document transmitted, not that such a document was genuine (i.e., had not been tampered with). Moreover, a facsimile can
never be considered a "copy in the form of original" (i.e., an original
lacking the quality of primitiveness) and consequently have the
same weight as an original; and not only because it lacks some of
the extrinsic elements of the document transmitted. In fact, an
original is the first complete and effective document, that is, an
original must present the qualities of primitiveness, completeness
and effectiveness. With facsimile transmission, the first two qualities belong in the document transmitted while the latter belongs in
the document received. This implies that the two documents to-

---

present throughout the centuries in all paper-based correspondence. While in the
ancient past a first draft (prepared by the sender) resulted in a final draft with
insertion of formulas, dates, etc. (prepared and maintained by the chancery, that is,
by the system), and in an original containing all the elements necessary to make it
effective (received and maintained by the addressee, and often maintained, in a
complete but not effective form, in a copybook or a registry, also by the sender), in
more recent times, a draft guide letter (handed by the sender to the records office
with a list of addressees and instructions to produce individual letters to send via
registered mail) resulted in many individual original letters (received and maintained by the various addressees), in as many copies (made and maintained by the
records office, that is, the system), and as many delivery acknowledgements (received and maintained by the system). If today, using an electronic mail system, one
drafts a message which may or may not have complex formatting, names a list of
addressees, and requires an acknowledgement of receipt, the documents sent and
received are not substantially different from those sent and received in paper-based
communication. The method of transmission has no influence on their nature nor
on their form: rather, what has an influence is the medium on which they are stored
to produce results. Thus, if one receives an E-Mail message and prints it out to
include it in a paper file relating to the transaction in question, one is in fact dealing
with a paper record containing information about its delivery, just like so many other
paper records.
    11. P. 48, n. 32.

gether constitute the original document, that they support each other, and that neither can be considered primary evidence on its own. The weight of a facsimile as evidence will remain subject to verification that it was sent, received and maintained in the regular course of business, as demonstrated by the "hearsay rule" (see, for example, the *Evidence Act*, R.S.B.C. 1979, c. 116, s.48 (1)), which advises against the rather common habit of copying facsimiles on archival quality paper for preservation purposes (either in the creating office or in an archives), and destroying those on thermal paper without going through a formal certification process (a process similar to that carried out for microfilm). Also, this rule places in a position of liability those who destroy the document which was transmitted and those who act on a facsimile when the transmitted document is disposed of. In fact, while a verification of the date of transmission and of the machines involved in the transmission is made easy by modern technology, that of the identity of the sender may be determined by established office routine (e.g., a log of those who use the fax machine), and that of the context in which the document was received by the addressee's record-keeping system; verification that the document received has the same content as the document that the sender claims to have transmitted is only possible by comparing the two documents.[12]

The above discussion does more than demonstrate that a significant contribution of diplomatics to archival thinking is the strict connection that it establishes between archival documents and the juridical system in which they are created. It shows how important it is for archivists to be able to identify the status of transmission of a document. It is often argued that because they deal with groups of documents rather than with single documents, modern archivists should not be concerned with originality. This statement is correct if its only intended meaning is that the archivist preserves archival material having different status of transmission, not just originals. But this does not, and should not be taken to, imply that the status of transmission of documents need not be considered by

12. For a complete discussion of facsimile documents see Erwin Wodarczak, "'The Facts about Fax': Facsimile Transmission and Archives" (M.A.S. Thesis, The University of British Columbia, 1991).

the archivist while accomplishing every archival function. If it is true that each file contains originals of the documents received and drafts and/or copies of documents sent, the archivist will have to be certain that this is really so, that the "file" under consideration is the original file (i.e., the first, complete and effective file), not a copy of it.

With electronically produced records in particular, the grounds for dismissing the status of transmission of a document as a relevant issue is that electronic records are always copies: "the archivist's concern should be that the documents in his/her possession are authentic and accurately reflect what the juridical person created at the time of the action."[13] It is certainly important to establish whether the records the archivist acquires are genuine, that is, are those made and received by the records creator in the usual and ordinary course of business. However, it is equally important to establish their status of transmission. In fact, electronic records are not *always* copies, because a copy is by definition a reproduction of an original, a draft or another copy (the first copy made being always a reproduction of a document in a different status of transmission); therefore, electronic records having a different status of transmission must be created for copies to exist. It is more appropriate to say that electronic records are all made as drafts and received as originals, in consideration of the fact that the records received contain elements automatically added by the system which are not included in the documents sent, and which make them complete and effective. Also, an electronic document comes into existence as a draft when its maker decides to save it for the first time, and comes into existence as an original when its addressee decides to save it for the first time, because information which is not affixed to a medium is not a document. When a document is made not to be sent but to be available and produce consequences within the creating body, it would probably be saved in its original (primitive, complete and effective) form only. So far,

---

13. David Bearman, letter to the author, loc. cit. See also: David Bearman, "Impact of Electronic Records on Archival Theory," *Archives and Museum Informatics* 5, no. 2 (Summer 1991),p. 6, and the report by Charles Dollar to which Bearman's article refers, which is mentioned in note 5, above.

there is no substantive difference between electronically produced and paper records as to status of transmission. However, it is fair to say that electronically produced records are generally used and maintained in the status of copy. The electronic copy of a draft would simply be a subsequent identical draft. The electronic copy of an original, if made while the record is current, would be a "copy in the form of original," that is, complete and effective but lacking the quality of primitiveness. The electronic copy of a record which has exhausted its effectiveness, if made within the same system or by transfer to another system, but using the same software, would be an imitative copy, because it would be identical to the original; the purpose for making it would not be the accomplishment of the transaction in which the original was actively involved, but the preservation of the evidence of it: the fact of not being created in the usual and ordinary course of busines would diminish the authoritativeness of the copy. The electronic copy of a non-current record made by transfer to another system in a software-independent way would be a simple copy if the data structure, or physical (and partly intellectual) form of the original were not explicitly captured. All this should demonstrate that it is important to ascertain (1) whether the electronic record copies we encounter at any given time are copies of drafts, originals or other copies, and (2) when they were made with respect to the actions and transactions in which the drafts or originals they reproduce were involved. These factors impinge on the record copies' effectiveness, authority, weight, authenticity and, possibly, on their genuineness.

When the archivist has to understand the relationships among interacting records creators, an important clue is the status of transmission of the respective documents. When appraising for legal or intrinsic value, deciding on conservation issues, 'migrating' electronically produced documents, even when arranging and describing groups of documents, archivists must consider, at every step, the status of transmission of the documents they are dealing with and its implications. As professionals whose primary functions are the "identification" and "communication" of documents,[14] archivists must be rigorous in their use of terminology; be

_____

14. This exemplification of the archivist's functions may seem excessive, but, on reflection, it is clear that appraisal for acquisition and selection, arrangement,

certain that the terms they adopt reflect the substantive nature of the entity they name; and, with respect to those terms which, notwithstanding their ambiguity, are commonly used, be aware of what they are referring to at any given time. For example, when legal or official records are defined as "recorded transactions,"[15] the intended meaning of the definition is all contained in the word "transaction." Usually, to a computer scientist and to archivists who have dealt mainly with electronic information systems, a transaction is any form of communication with a store of information, such as a database. All too often, specialized language interferes with our work. Archivists have to keep in mind that records are created in an administrative context, within a juridical system, not within the limited boundaries of a specialized discipline or technology; therefore, the terms that they apply to the records must have the meaning given to them by the administrative-juridical context of the records themselves. Diplomatics recognizes this contextual relationship and emphasizes it. In the case of the word "transaction," for example, the diplomatic definition is "an act or several interconnected acts in which more than one person is involved and by which the relations of those persons are altered." This means that to have a transaction it is not sufficient to have a communication, but it is necessary that such a communication creates, modifies, maintains or extinguishes a relationship with other persons. If it is done only for the purpose of viewing information, accessing a database is not a transaction and does not produce official records; rather, it is a mere act the result of which only relates to consciousness. If instead a database is accessed as part of the process of carrying out a transaction, this action may produce official records. To have an archival document, it is necessary to have an action manifested in writing for the purpose of bridging space and/or time, that is, it is necessary to have a communication with other persons or with oneself. While only the

---

description and reference, consist mainly of identification of documents and their context for purposes of communication.

15. Pp. 67-70, 73-75.

former type of communication can be part of a transaction and generate official records which participate in a procedure, the latter type can only be an action and generate non-official records, or documents of process, such as those usually called "working papers," diaries, personal agendas or notes, or ... documents of access to information systems. With regard to artificial intelligence, if the juridical system recognizes an expert system, for example, as a person, that is, as an entity capable of acting legally, then communications with it are to be considered transactions.

However, terminological rigour is not the only way in which diplomatics contributes to functional consistency and allows for meaningful standardization.[16] It is often stated that archivists must become directly involved in the process of creating archival documents, and specifically in the design of electronic information systems and the definition of the standards governing those systems. To do so, archivists must be able to see the archival document primarily as embodiment and evidence of action. Diplomatics makes explicit the links between the intellectual components of a document and the elements of a typical act, and in so doing facilitates the determination of a documents profile, just as knowledge of structural grammar facilitates the composition of a text and makes it understandable to the reader.

Diplomatics also emphasizes the relationships among documentary forms, types of acts, and procedural phases, and shows all the types of interaction between persons and documents. A clear understanding of such relationships and interactions enables the archivist to advise records creators on what Schellenberg called simplification of functions, work processes, and records procedures, and considered to be the foundation of any records management activity;[17] and, with respect to electronic information systems, to advise records creators on capturing information about

---

16. It is well known to the readers of this book that the definitional component of diplomatics does not encompass only the elements discussed here as examples, but also the persons concurring in the formation of a document, the acts represented in it, the procedures which determined its creation, and its internal components.

17. Theodore R. Schellenberg, *Modern Archives. Principles and Techniques* (Chicago, 1956. Midway Reprint 1975), pp. 44-51.

their systems in "metadata systems" which document the input and output products, the relationship among files, the nature of software facilities, and the functions supported by the systems.

However, more than anything else, this understanding serves to balance the modern societal trend towards information and away from documents. Now more than ever, the first responsibility of the archivist is the preservation of the nature of archival documents as means for action, of their evidential quality, of their ability to perpetuate the deeds of our society.[18] The concrete record with its stable form and direct relationship to activity stands in contrast to the intangibility of the concept of information. "Records are at the 'still point of the turning world' possessing 'neither arrest nor movement' to use some lines from an archivist/records manager's favourite quartet of poems (T.S. Eliot—*Burnt Norton*—one of the *Four Quartets*). Records have evidential value precisely because they have an element of stability."[19]

What about "virtual documents"?—more than one archivist has posed this question. Virtual documents are documents of process; they are directly related to actions and have established internal relationships, just like any first draft of a document of any type; they stay with their creator in their virtual form, but, if communicated to other persons, they reach the addressee(s) in a complete and effective form, as records of a transaction.[20] Virtual documents are archival documents, not representations of facts and figures meant for processing and interpretation, such as, for example, the

---

18. More and more contemporary archivists are beginning to take a position against the view that records are just a subset of an agency's information system—and not only in North America. See for example the articles by Frank Upward ("Challenges to traditional archival theory"), Michael Sadlier ("Plus ça change ... or Forward to the past or Sir Hilary triumphant"), and Glenda Acland ("Archivist—Keeper, undertaker or author: the challenge for traditional archival theory and practice"), in *Keeping Data. Papers from a Workshop on Appraising Computer-Based Records* presented by the Australian Council of Archives and The Australian Society of Archivists Incorporated on 10-12 October 1990, edited by Barbara Reed and David Roberts (Sydney, 1991), pp. 105-119.

19. Frank Upward, "Challenges," p. 106.

20. Virtual documents are remarkably similar to the "imbreviatures" of medieval notaries. A notary who was asked to prepare a document would take from the first party a blank parchment, fold down its top right corner, and annotate on it the type of action the document was meant to put into existence, the names of the persons involved, dates, terms of settlement, and other details. The document was main-

data in a Geographic Information System. Diplomatics provides the concepts and the principles necessary to clarify these distinctions in the mind of the archivist, because, just like transformational grammar, at a well defined point it moves the emphasis from the analysis of documentary components to the way persons interact by means of documents. But does diplomatics also provide a method for accomplishing archival functions?

## Diplomatics as a Method of Inquiry

> From our personal experience in the arrangement of archival fonds prior to 1940, [and] from a certain number of problems encountered in the course of the appraisal of contemporary archives, ... we have derived ... the conviction that it is not possible to manage archives if they have not been identified and analysed.
>
> Gérard and Christiane Naud[21]

Any scientist would appreciate the need to study the elements in order to understand the cosmos. That is axiomatic. Just as the scientist strives to understand the elemental constitution of the physical world, so do we with regard to the object of our work, the archives. Diplomatics offers the instruments for gaining such an understanding. However, some modern archivists consider diplomatics to be reductionist or atomistic; forgetful of the global, holistic aspects of human activity and of the archival axiom that the whole is much greater than the sum of its parts. They point out that modern archivists work almost exclusively with aggregations of

---

tained in this form by the notary who, if needed, would later prepare the original by adding standardized formulas taken from a "formularium" to the data. The original would stay with one of the parties, or both of them (as in the case of an indenture or a duplicate original).

21. "De notre expérience personnelle dans le classement des fonds d'archives antérieurs à 1940, [et] d'un certain nombre de problèmes rencontrés lors du tri des archives contemporaines, ... nous tirons ... la conviction qu'aucune gestion d'archives n'est possible sans que celles-ci aient été identifiées et analysées." Gérard et Christiane Naud, "L'analyse des archives administratives contemporaines," *Gazette des Archives* 115 (4e trimestre 1981), p. 216.

documents, that it is misleading and essentially impossible to start our description of a fonds with the analysis of the documents, and that we have rather to go from the functions to the records, from the general to the specific. The main argument of these archivists is that the historical/administrative/juridical context of the document is almost always *known* or *knowable*, and that diplomatics may offer clarification but not a method of analysis.[22] The words at issue here are those emphasized: "known" and "knowable."

How do we acquire a knowledge of the provenance of a fonds? Statements of mission, mandates, legislation, regulations, official reports, organizational charts, internal circulars, and other similar documents are usually indicated as the archivist's main sources. But we know that a continuous mediation takes place between the legal/administrative apparatus and its human component, and that law and administration have a natural inertia and adjust to change long after it has taken place. Even more importantly, we know that function and competence influence the content of archival fonds, and that organization influences its overall structure; but it is the actions and transactions, and the procedures by means of which they are carried out, that determine the form of the documents, their interrelationships, and their quantity. It is essential to recognize how the informational content of the archival fonds is determined by the functions of its creator, how its shape (the organization of collectives of documents within the fonds) is determined by the organizational structure within which it was produced, and how the form and interrelationships of its records (within each collective) are determined by the activities and procedures which generated them.

Even if the sources which archivists normally use to acquire knowledge of the functions, capabilities and organizational structure of a records-creating body were entirely reliable, the knowledge we could obtain from them would be limited to the kind of informational content we may expect to find in the records of any

---

22. See, for example, Terry Cook, "Mind Over Matter: Towards a New Theory of Archival Appraisal," in the *Festschrift* in honour of Hugh Taylor (Association of Canadian Archivists: Ottawa, 1992. This type of scepticism as to the usefulness of diplomatics as a method of analysis is usually presented with regard to the records of large organizations.

given agency. This knowledge would not only be hypothetical, but also grossly insufficient for making any kind of archival decision. It would be necessary to know the specific activities of each agency, but, while looking for them, one would soon discover that "all activities are to be brought back to procedures," because every activity follows a certain pattern in passing through certain well defined steps, and those procedures are fully revealed only by the form of the records.[23] Furthermore, one would discover that record forms correspond to informational content, that is, each record form typically carries a certain type of information and is linked to the other record forms by a well defined kind of relationship. This is the reason why Dutch archivists have embarked on the very complex endeavour of examining single record forms independently of specific organizational contexts, and writing a commentary for each of them which details when and where it appeared for the first time; why it was created; what its purpose was with regard to the type of activity generating it; what kind of informational content it held; how it looked; how it changed over time as to appearance and purpose; why it disappeared/if it did disappear; to what other record forms it was and/or is usually connected by functional and procedural relationships; and finally, which type of juridical person normally creates it. This study is conducted not only for record forms which pertain to isolated documents, but also for forms of record aggregations, such as various types of registries, volumes and, supposedly, files.[24]

When work of this kind is completed for all recurrent (as opposed to unique) forms of archival material, this knowledge will inform how current records are described and listed in records

<hr>

23. This point was clearly made by Peter Sigmond in his paper "Form, Function and Archival Value: The Use of Structure, Forms and Functions for Appraisal, Control and Reference," presented at the Annual Conference of the Association of Canadian Archivists (Banff, Alberta, 25 May 1991).

24. See Vereniging van Archivarissen in Nederland [Association of Archivists in the Netherlands], *Handleiding by het vervaardigen van een broncommentaar [Manual for the preparation of a source commentary]*. (Gravenhage-Nijmegen, 1987). Unpublished English translation by Hugo L.P. Stibbe, "unverified, and done for pleasure for whomever may be interested in this work by Dutch archivists." See also David Bearman and Peter Sigmond, "Exploration of Form of Material."

inventories, classification schemes, retention and disposal schedules, metadata systems and indexes, because all of these will pre-
sumably be based on controlled vocabularies and authority files for
terms referring to actions, procedures, and forms. To a degree,
archivists will then not need to see the records to know what they
are all about, because terms, when linked to procedures and actions, can communicate content.[25]

It may be argued that with electronic records, the relationship
between record forms and procedures, and between record forms
and the type of information they carry, are still evolving.

> We do not have a comprehensive or stable set of categories for
> different types of electronic records forms, nor have we established
> clearly the relationships between certain forms and content or pro-
> cedures. In my view, there are two reasons for this. One is that these
> relationships are changing (largely through trial and error). The
> other is that stable forms of electronic records have not been subject
> to thorough analysis.[26]

This is true, but does not diminish the relevance of diplomatics
with respect to electronic records. In the first chapter, it was stated
that where there are not rules governing the creation of records, a
knowledge of diplomatic principles and concepts "gives those who
try to formulate those rules a clear indication of the elements which
are significant and must be developed." Briefly, where records
creation is consciously controlled, diplomatics guides the recognition of patterns and facilitates identification, while, where records
creation is uncontrolled, diplomatics guides the establishment of
patterns, the formation of a system in which categories of records
forms are devised, which is able to convey content and reveal
procedure. Once a system is established, then its description in a
metadata system will have to reflect it by expressly articulating the
relationships among record forms, procedures, actions, persons,
functions, and administrative structures.

---

25. Sigmond, "Form, Function, and Archival Value." Of course, archivists will
still need to see the records for purposes of control.
26. Margaret Hedstrom, letter to the author, 4 November 1991.

The main problem identified by Gérard and Christiane Naud is that the various record classification schemes, metadata systems and records descriptions tend to confuse the actions generating the records with the subjects of the records.[27] Naud and Naud believe that appraisal is difficult because classifications, and generally any description of current records, fail to adequately represent them, and that those descriptions fail because identification is based on inconsistent and inappropriate criteria. As a consequence, the archivist entrusted with appraisal is often in need of reanalysing the records. To do so, the top-down approach, that is, an approach which begins with an understanding of the creator by means of acquiring knowledge of its functions and organization on the basis of laws, regulations, and the like, is a useful starting point; but, as Heather MacNeil asserts it, it

> should properly be viewed as a supplement to, not a replacement for, the more traditional bottom-up approach. The illumination of the provenancial and documentary relationships that are embodied in organizational structures and bureaucratic procedures, and embedded in documentary forms, depends upon an analysis that continually mediates between acts and the documents that result from them. These relationships can only be brought into unconcealment with the simultaneous application of a bottom-up analysis, which is most clearly typified by the diplomatic analysis of the genesis, forms, and transmission of documents. Such analysis is critical to ensure that the documents we bring into archival custody actually reflect, accurately and meaningfully, the functions, activities, transactions and rules of procedure that shaped their formation; in other words, that they do what they are supposed to do.[28]

---

27. Naud, "L'analyse des archives administratives contemporaines," pp. 221-225.

28. Heather MacNeil, "Commentary on Peter Sigmond's 'Form, Function and Archival Value: The Use of Structure, Forms and Functions for Appraisal, Control and Reference'" pp. 11-12. David Bearnman agrees with MacNeil's approach. He believes that organizational and diplomatic analysis must complement each other, "unless only one source of information is available." In electronic systems—he writes—transactions are always products of applications. Only if designers make applications so that they have the same boundaries as organizational business purposes, can the two types of analysis become one. (letter, loc. cit.).

These words clearly explain the contribution of diplomatics to the analysis of archival material. Diplomatics gives importance to the broad context of creation by emphasizing the significance of the juridical system (that is, the social body plus the system of rules which constitute the context of the records), the persons creating the records, and the concepts of function, competence, and responsibility; but never distances itself from the reality of the records. Furthermore, the diplomatic axiom that record forms convey and reveal content is essential to the formation of the missing link between the provenance and the pertinence approaches. The principle of provenance, as applied to appraisal, leads us to evaluate records on the basis of the importance of the creator's mandate and functions, and fosters the use of a hierarchical method, a "top-down" approach, which has proved to be unsatisfactory because it excludes the "powerless transactions," which might throw light on the broader social context, from the permanent record of society. This difficulty has opened the door to the principle of pertinence which, as aplied to appraisal, enables us to evaluate records on the basis of the matters to which they pertain. Sometime pertinence is viewed as the umbrella within which a provenancial approach can be used: first, relevant topics or issues are defined; then the records creators involved with those topics and issues are identified; and finally, their records are evaluated according to provenance. Sometimes pertinence is viewed as a subsidiary approach which could compensate for the shortcomings of a pure provenancial one, and is subsumed under provenance: first, significant creators are identified, then the important topics or issues they deal with are defined, and finally the records are evaluated according to their subject matter. At other times, pertinence is viewed as a pure method of appraisal. In all cases, this approach is invariably criticized as bieng extremely subjective. The real problem is that provenance and pertinence have been transported from the realm of arrangement to that of appraisal, as two antagonistic "principles," and have been used *de facto* as two alternative "methods," while they are extraneous to appraisal both as principles and as methods. However, they are relevant to appraisal as "conceptual goals": the societal records must reflect their creators' mandates, functions and activities, *and* the societal interplay, the societal fabric, the main events and issues of each era. The method for reaching this goal must find its theoretical foundation in the nature of the records themselves. Their nature, having been determined by the circumstances of creation, imposes a contextual and analytical approach, and requires that appraisal be conducted by examining the records' creators' functions, activities and procedures, *and* the relationships between these and the records created, and among the records themselves. But in order to understand these relationships we have to focus on the records themselves and on their form and genesis, that is we need a "bottom up" approach which complements the "top-down" approach. At this point, provenance and pertinence goals can be reached by means of the contextual method. Therefore, the use of diplomatic analysis has the capacity of eliminating the dicholtomy between provenance and pertinence, and providing the channel which allows one to flow naturally into the other. This is to be remembered not only when we conduct appraisal for the selection of records within fonds which our archival institution must acquire by mandate, but also first and foremost, when we undertake appraisal for the acquisition of private archival materials in order to complement our institutional holdings, or to fulfill the mission of a thematic archival repository.

This is not the proper place to discuss the relative merits of the various approaches to acquisition. Whatever the method we use to determine the acquisitions policy of an institution, be it focused on a geographical area, a legal jurisdiction, a type of records creator or a subject area, once that policy is developed, we need a systematic approach to implementing it. Locating records dispersed among the relevant records creators, identifying the "documentary problems,"[29] and which records to acquire, necessitate an analysis much deeper than the study of published literature and official sources. It is quite obvious that we need that type of information for our initial inquiry, in order to understand the types of records creators, their history and characteristics, their general purpose, and even their functions. However, this can bring us only as far as the identification of some records creators who *may* have the records necessary to fulfill our mandate. We still have to *see* those records

---

29. This expression is often used by Helen Samuels in her writings on documentation strategy.

and understand how they are procedurally and formally interrelated. This is not a "records survey." Rather it is a "records analysis," which may guide us to other, more significant, records creators who are "procedurally" linked to the activities generating the records under examination.

For example, let us consider the copies of clinical studies held by a pharmaceutical company. These may tell us that the originals are in the hands of the research group responsible for those studies; their documentary form can tell us the procedure followed in the research, and guide us to drug tests the existence of which we would not have suspected; their relationship with the procedure (e.g., they are part of the deliberation phase rather than the execution phase) can reveal how responsibilities were distributed and again guide us to complementary records; whether that procedure was instrumental or constitutive can suggest the relative value of the records we encounter; and the intrinsic elements of form can show us whether we are confronted with routine documents or with a special dossier. We must let the records speak for themselves, not by reading their content or calculating their extent, or looking at their classification or preparing lists, but by analysing their forms, formation and relationships. These other activities may come after we have decided whether to acquire and what to acquire, and this we can decide only when we have gained an understanding of how the actions and transactions of the records creator resulted in records, and how those records relate to others within the same fonds and to those in the fonds of different records creators. It is not by reading regulations and identifying functions that we discover the societal interplay; what determines that interplay are the transactions, that is, the actions in which more than one person is involved and by which the relations of those persons are altered. Although the general study that we conduct at the outset can provide us with more than mere clues, it is the reality of the records themselves which provides us with knowledge. To say that our analysis must start from the records does not mean that we can neglect the prior historical/juridical/administrative research; it only means that *archival description proper* starts from the records, after the *historical-juridical* work has prepared the ground for it. And proper archival research, conducted on "innocent and impartial" sources, may uncover a reality which is inconsistent with that revealed by previous research.[30] As well, archival work may be hampered by our understanding of the reality based on that research: expecting to find certain things, we may have trouble recognizing the different things that we actually see, and our eyes may be so blinded by our preconceptions that we may try to constrain the reality to fit our hypothesis.[31] Therefore, our initial study should be just a beginning, and no more.

After all is said and done, appraisal, either for acquisition or for selection, is "a work of careful analysis."[32] As Barbara Craig puts it, we need "a more sympathetic orientation to records where respect replaces control as the basis for decision."[33] And she adds

> It seems to me that we need not just documentation plans, nor plans with the addition of administrative context, but plans, context and a knowledge of documents, records and their forms. The reality of the record base is an indispensable component of all acts of appraisal. Without an understanding of documents and records, of their forms and of their functions and of how they were created and used, plans can easily be divorced from reality .... it will be a sad day and a dangerous step when faith in planning replaces the study and knowledge of records.[34]

If the most controversial service of diplomatics is that which it provides to appraisal, the most obvious and accepted is that which it provides to arrangement and description. The capacity of diplomatic analysis to uncover the interrelationships of records makes

---

30. Records are innocent and impartial sources because they are created as a means for action, not as a purpose in themselves. Naturally they contain the biases and idiosyncrasies of their creators, but, because they were not meant for dissemination, they have the capacity to reveal what really happened.

31. The author speaks from personal experience. When examining the supposed records of the French Government in Rome after having conducted a complete historical-juridical and functional study on the subject, I wasted a great amount of time trying to make the records fit into my preconceptions, and might never have seen the truth if I had limited myself to an identification of the structure of the fonds (see p. 97 n. 13).

32. Terry Cook, *The Archival Appraisal of Records Containing Personal Information*, p. 38 [ms]. Before Cook, Schellenberg, Brichford and others have stressed this point.

33. Craig, "The Acts of the Appraisers," p. 7.

34. Ibid., p. 11.

it a precious instrument for arrangement, be it physical or purely intellectual. Its conceptual and terminological rigour allows for the proper identification, naming and formulation of the data elements to be entered into a description. Moreover, diplomatics presents the records universe as "one capable of being broken down analytically and compartmentalized into its constituent elements," that is, one "eminently amenable to standardization." Standardization of archival description tries to do exactly what can be accomplished by applying diplomatics: to extract from what we observe the elements which are relevant, and to name them; to define those elements "in a way that clearly differentiates information pertinent to creators and their functions and activities" and procedures, from that pertinent "to the records created out of those functions and activities" and procedures; "and, finally, to organize those elements in a logical order."[35] The process necessary to carry out this operation can also generate controlled vocabularies, authority files, and other similar instruments, which allow researchers "to act across jurisdictional boundaries and structural accidents to identify commonalities of human action in [the] administrative environment over time."[36] David Bearman writes that

> Research into the nature of documents is now demonstrating not only that 'document formalisms' (structural features of records that signal their contents to the culturally attuned) exist, but also that machines can be taught to distinguish between document types. Computers can parse documents for their internal components and 'mark' them with such document-marking languages as Standard Generalized Markup Language (SGML), creating a sort of electronic 'fingerprint' of a form-of-material... these files... look... like records schedules without dates or names of offices; they contain a field for SGML-like 'fingerprints' and fields for data elements typically found recorded in this type of record... Such fingerprints might also form the controlled vocabularies that link reference files of document types to databases of archival records.[37]

---

35. MacNeil, "Commentary...", p. 8.

36. David Bearman, "Archives and Manuscripts Control with Bibliographic Utilities," *The American Archivist* 52 (Winter 1989), p. 33.

37. David Bearman, "Authority Control Issues and Prospects," *The American Archivist* 52 (Summer 1989), p. 298.

The above discussion should demonstrate that diplomatic methodology is not a substitute for collective archival processing, but an analytical method of inquiry. While diplomatic principles constitute a fundamental instrument for understanding the object of our professional responsibility—the record—diplomatic methodology is a means for learning about documentary, administrative and juridical context, a context which is the focus of all the archivist's functions. If diplomatics assists us in providing advice to records creators and systems designers, and in appraising, arranging, describing, and communicating the documentary products of societal endeavour, it does so as a mediator which makes explicit for us the elements on which to base our decisions and assess their relative value.

One may wonder whether, in carrying out their functions, European archivists set aside a specific phase of their work for conducting a diplomatic examination of the records. Of course, they do not. They are as unaware of using diplomatics in the course of their analysis of the records as a reader is unaware of using syntax while reading a book. However, a reader acquires such awareness when he or she is confronted with a book in a foreign language. And a fonds no part of which we have ever acquired before is just like a foreign language to us: we may already know its origin and cultural foundations, and much of its vocabulary if we have been exposed long enough to it, but we need to learn its grammar and syntax to be able to understand it, and we can do so better if we already know the fundamental concepts of grammar and syntax: that is, what is a noun, an adjective and a verb; the difference between indicative and subjunctive mood; between subject and complement; a hypothetical and conditional sentence, etc. But, even after we have learned the specific grammar and syntax of the language, we may have to stop and reflect on its rules every time we encounter a particularly complex paragraph.

This book was meant to introduce North American archivists to a very old discipline which has the potential of guiding their work in the most unpredictable ways, the discipline of the records. While illustrating diplomatics, this book has focused on the nature of archival documents as determined by the circumstances of their creation and as revealed by their forms. By doing so, it has emphasized context over content, purpose over use, and it has posed some of the most fundamental questions which must be asked in order to gain an understanding of archival materials.

The archival document, given its stability and concreteness, occupies a central place both in our professional knowledge and in our work. Jay Atherton's concept of continuum *versus* life cycle, implying an integrated approach to records and archives management, that is, to all archival functions,[38] makes sense only on the assumption that (1) the reality of the material with which we work is determined by the juridical system in which it was created, by the persons concurring in its formation and their competence, and by the actions, transactions, processes, and procedures which generated it; and that (2) once established, such a reality is immutable in its form as well as in its substance.

Human ingenuity will continue to devise new types of documents. It is conceivable that one day we shall access the information maintained by our planetary system by looking at the sky with a telescope. Shall we be able to recognize what part of it is the record of our society and protect its integrity, which after all is what every user of archives should expect of us? The answer will depend on our ability to identify the enduring substantial components of a record. Perhaps they will be the same as those established by Dom Jean Mabillon in 1681 in his *De Re Diplomatica Libri VI*.

---

38. Jay Atherton, "From Life Cycle to Continuum: Some Thoughts on the Records Management-Archives Relationship," *Archivaria* 21 (Winter 1985-86), pp. 43-51.

# Index